T0210820

Unfolding Meaning

'... to think differently – this thought must enter deeply into our
intentions, actions, and so on – our whole being.'
David Bohm in 'The Implicate Order:
a new approach to reality'

In *Unfolding Meaning*, David Bohm, one of the most provocative
and original thinkers of our time, argues that there are other ways of
thinking to bring about a different, more harmonious reality. Our
fragmented, mechanistic notion of order derives from the modern
conception that our earth is only part, not – as it was with the Greeks
– the centre, of the immense universe of material bodies. The
implications of this idea permeate modern science and technology
today and also our general attitude to life.

The dialogue develops as an attempt to find another way of
thinking; it is an exercise in unfolding some of the vagaries of
thought, by forty-four peole who gathered to meet with Professor
Bohm to consider with him some of his ideas on a number of
subjects: from implicate order to soma-significance, from fragmentation to wholeness.

The late **David Bohm** was Emeritus Professor at Birkbeck
College and a fellow of the Royal Society. His work in physics had
led him to propose the idea of quantum potential, a means by which
the view of universal, unbroken wholeness, implicit in relativity
theory, might be understood in the context of the more abstract,
fragmentary approach of quantum mechanics. Moreover, he was
interested in the philosophical implications of quantum and relativity
physics. He wrote many books including *Wholeness and the
Implicate Order* and *Thought as a System* (both published by
Routledge).

Unfolding Meaning

A Weekend of Dialogue with David Bohm

David Bohm

Routledge
Taylor & Francis Group

LONDON AND NEW YORK

First published 1985
Ark edition 1987
by Routledge
2 Park Square, Milton Park, Abingdon, Oxon, OX14 4RN

Simultaneously published in the USA and Canada
by Routledge
605 Third Avenue, New York, NY 10017

*Routledge is an imprint of the Taylor & Francis Group,
an informa business*

Notice:
Product or corporate names may be trademarks or registered
trademarks, and are used only for identification and explanation
without intent to infringe.

British Library Cataloguing in Publication Data
A catalogue record for this book is available from the British Library

Library of Congress Cataloging in Publication Data
A catalog record for this book is available from the Library of
Congress

ISBN 13: 978-0-415-13638-9 (pbk)

CONTENTS

ACKNOWLEDGEMENTS

Teachers often comment that they learn as much from their students as they actually impart, but seldom does this seems to be the case from the point of view of the student. Occasionally though, a flow of increased understanding does take place between teacher and student. This tends to happen when the separation between those roles breaks down and a flow of dialogue imbued with mutual respect takes place. Then a true collaboration occurs, and the result is something greater than might have come from the more usual, simple transference of information.

Professor David Bohm is one of those rare men who recognizes and enjoys this way of working — or, as he would put it, participating in a dance of the mind. It has been a great privilege to have had the opportunity of sharing his thoughts and his company.

I would like to take this opportunity to thank his wife Sarah Bohm whose enthusiastic encouragement helped to make the event upon which this book is based, such an enriching experience. I would also like to thank Peter Garrett, the European co-ordinator of The Foundation of Universal Unity, (now the Emissary Foundation International) for organizing the event; the staff and management of the Three Ways Hotel for their remarkable care and skill in providing a setting that facilitated the smooth flow of our deliberations; Cliff Penwell, Mynda Itzigsohn, and especially Lesley Wilson and Tuli Corbyn for their efforts in helping to transcribe the very complicated conversations on the taped recordings, and Lindsay Rawlings for his contribution of the cover design and photograph for this edition.

Also David Bohm wishes to acknowledge the immense value of earlier discussions with J. Krishnamurti and with Dr. P. de Mare among others.

Donald Factor

INTRODUCTION

by

Donald Factor

Ideas, concepts and theories are the stuff of thought, and thought affects the world in pervasive ways. What we think about reality can alter our relationship to it, just as what we perceive in the world around us can alter our thoughts. Thought is the ground upon which our understanding rests. With thought we see the world and in a continuing process learn to interact with that world. We can look beyond our raw perceptions and alter the course of our actions. We can solve problems; we can create new products, technologies, ways of dealing with our environment and with one another.

But much of what we think remains hidden from our conscious awareness. Within our minds we carry a record of past experience, of lessons learned, of incidents and details long forgotten. Our thoughts are coloured and conditioned by such limits as our language and our culture. We interpret our experience through a mixture of conscious and unconscious memories, imaginings and desires, and with these we organize our world. Often our thoughts, when acted upon, lead to unexpected and sometimes unimagined results. They seem to contain unrecognized implications of meaning of which we knew nothing, and that appear in spite of what we might have thought was our complete understanding. How then might we evaluate our thought? How might we discover whether or not our most cherished ideas are in fact valid and relevant to the circumstance before us? What do our thoughts mean?

This book is a record of an experiment in unfolding some of the vagaries of thought — an experiment conceived and developed during the course of a weekend of conversation between forty four people who gathered to meet with Professor David Bohm and to consider with him some of his ideas on a far-ranging list of subjects. All had some familiarity with his work and an interest in looking further into its implications. Many had attended various conferences, seminars and workshops where a leader, or an invited expert, either taught or guided the participants toward an increased understanding of his or her area of expertise. This weekend turned out to be very different.

David Bohm is Emeritus Professor of Theoretical Physics at Birkbeck College, The University of London. His work in physics has been predominantly concerned with the problem of motion and process which relativity physics deals with but quantum theory does not. Out of this interest he proposed the idea of a quantum potential, a means by which the view of universal, unbroken wholeness, implicit in relativity theory, might be understood in the context of the more abstract, fragmentary approach of much of quantum mechanics. His theory of the implicate order, an approach whereby implicit potentials can be seen to unfold out of a universal, unbroken field into explicit phenomena before being re-enfolded, has provided a new and valuable interpretation of quantum mechanics, and has provided a basis not only for new insights in physics but also for a whole range of other subjects.

For many years Professor Bohm has been especially interested in the philosophical implications of quantum and relativity physics, and with discovering a metaphor that might make their meanings accessible to a general public unfamiliar with the mysteries of higher mathematics. His feeling has been that this is important because the mechanistic world-view that seems to dominate contemporary science and society has led to a state of increasing fragmentation, both within the experience of individual human beings and in society as a whole. The fact that this world-view is incomplete, and that it has not been widely recognized as such, has caused it to become bound up within a broad area of misunderstanding deriving largely from a misunderstanding of science in general, but also — and more importantly — from a general confusion regarding the nature of thought and of its relationship to reality.

He has suggested that thought is, by nature, incomplete. Any thought, any idea, any theory, is simply a way of seeing, a way of viewing an object from a particular vantage point. It may be useful, but that usefulness is dependent upon particular circumstances — the time, the place, the conditions to which it is applied. If our thoughts are taken to be final, to include all possibilities, to be exact representations of reality, then eventually we run up against conditions where they become irrelevant. If we hold to them in spite of their irrelevance, we are forced either to ignore the facts or to apply some sort of force to make them fit. In either case fragmentation is the result.

Professor Bohm's writings on universal wholeness, and his proposals concerning the implicate order have begun to have an influence on diverse disciplines. His ideas are central to what has become known as 'the holographic paradigm'. These ideas, which are explained and discussed in the main text of this book, have provided a new way of understanding a great many phenomena ranging from some of the problems of quantum physics to health care, social organization, religion, and the workings of the human mind itself.

In order to provide an opportunity to inquire more deeply into some of his thoughts, The Foundation of Universal Unity invited Professor Bohm to spend a weekend discussing these thoughts with a group of people of varying ages, nationalities and professional background. The intention was to discover if, by careful attention, a new and more fruitful vision of the possibilities for a greater harmony in the individual and in society might arise.

On the 11th of May, 1984, the group gathered at a small, country hotel in the Cotswold village of Mickleton, Gloucestershire, England. Professor Bohm, accompanied by his wife Sarah, arrived seeming tired and preoccupied. This was to be his first experience of such a gathering. He had come prepared to give three talks, and to then develop his ideas with the group through question and answer sessions. As the weekend unfolded, though, a very different experience began to emerge both for Professor Bohm and for all the participants.

The sessions developed an atmosphere of contained, mutual concern for the revelation of deeper insights. A spirit of friendship and respect between all those present emerged, and this quickly grew into a harmonious field where proposals of many sorts could be collectively investigated in safety and allowed to expand into new levels of understanding. A dialogue developed in which each participant was able to put aside his own views and listen to those of others. It became increasingly clear that no point of view was in itself complete, and that a collective process of thought was the means by which understanding could be enriched. This fact became the focus of the group's attention. No conclusions were reached nor were any programs initiated; rather the appreciation of a continual unfoldment of new insights revealed through friendly conversation was seen to be the means by which an increase of harmony might appear.

When such a process is translated into print it tends to take on the appearance of a finished product. The atmosphere out of which it emerged disappears, leaving only the arguments by which the various speakers hope to win agreement. Abstracted from the context of their creation the ideas stand naked, vulnerable to judgement, to criticism, to mere acceptance or rejection. This of course, is one reason for preserving ideas in print. As Professor Bohm suggests in the course of these discussions, 'Ideas must be vulnerable.'

The ideas considered in these pages should be seen as part of a work in progress. They represent a slice of a creative process and they are presented not as conclusions but as an example of one way that new ideas might be raised, inquired into, and allowed to unfold further. They also introduce a new phase of Professor Bohm's work, one in which the interactions between a group of individuals provide the focus of energy in which new meanings might be perceived, and where in his terms both the content and context of thought enfold each other, and unfold into new meanings and insights.

In a conversation between forty five people there is much apparent clumsiness. People do not share their thoughts aloud in perfect sentences of a sort that the reader of a book might ordinarily demand. There are many false starts, many incomplete proposals. Often in the course of these sessions, questions were raised, or statements made, that seemed irrelevant; but just as often, these opened the way to new and deeper levels of understanding. In attempting to document the proceedings I have tried to preserve as much as possible the flavour of the event. I have opted for a balance that might make the ideas intelligible, while preserving something of the flow of interaction between the participants that was central to the experience. I have used the term 'Question' to mark the contributions by participants other than David Bohm, although only in the early stages of the dialogues did they particularly tend to take the form of questions. As the conversations progressed they became, simply, parts of the emerging whole.

I have only been able to include here the dialogues in which the entire group was present. In addition to these main discussions there were other sessions in which the larger group split into three smaller groups, and of course there were numerous more intimate conversations over meals, and so on.

THE IMPLICATE ORDER
A NEW APPROACH TO REALITY

Professor Bohm: Throughout history there has been a succession of world views; that is, general notions of cosmic order, and of the nature of reality as a whole. Each of these views has expressed the essential spirit of its time, and each of them in its turn, has had profound effects on the individual, and on society as a whole, not only physically, but also psychologically and ethically. These effects were multiple in nature, but among them, one of the most significant is notions of universal order.

I'll begin by giving two examples of world views that are of key importance in this discussion. The first of these is the ancient Greek notion of the earth at the centre of the universe, and the seven concentric spheres in the heavens in an order of the increasing perfection of their natures. Together with the earth, they comprised a totality that was regarded as an integral organism, with activities they regarded as meaningful.

As suggested, especially by Aristotle, each part had its proper place in this organism, and its activity was seen as an effort to move toward that proper place and to carry out its appropriate function. Man was thought to be of central importance in this whole system, and this implied that his proper behaviour was to be regarded as correspondingly necessary for the over-all harmony of the universe.

Now in contrast, in the modern view the earth is a mere grain of dust in an immense universe of material bodies — stars, galaxies, and so on — and these, in turn, are also constituted of atoms, molecules, and structures built out of them, as if they were parts of a universal machine. This machine, evidently, does not constitute a whole with meaning — at least, as far as can now be ascertained. Its basic order is that of independently existent parts interacting blindly through forces that they exert on each other.

The ultimate implications of this view of universal order are, of course, that man is basically insignificant. What he does has meaning only in so far as he can give it meaning in his own eyes, while the universe as a whole is basically indifferent to his aspirations, goals, moral and aesthetic values, and, indeed, to his ultimate fate. It is clear that these two views will, in the long run, have very different implications for our general attitude to life, which can be profound and far reaching. For example, man tends to feel much more at home with an organic point of view — organismic.

Toward the end of this talk I'll discuss some of these implications in more detail. But for the present I'll merely call attention to the fact that a mechanistic notion of order has come to permeate most of modern science and technology, and for this reason has begun to affect the whole of life.

Now it's in physics that the mechanistic world-view obtained its most complete development, especially during the nineteenth century when its triumph seemed almost complete. From physics, mechanism has spread into other sciences and into almost all fields of human endeavor — that is, the mechanistic attitude. So some examination of the form that mechanism has taken in physics is called for if we are to understand what has by now become a more-or-less dominant world view which deeply affects all of us. In this examination the correctness and necessity of mechanism has to be evaluated and criticized, especially with regard to whether or not the actual state of knowledge in physics continues to sustain and support this view, as well as to whether or not alternative views are possible.

I'll begin by listing the principal characteristics of mechanism to make this idea more clear, and contrast its main features with those of an organismic type. Now firstly, the world is reduced as far as possible to a set of basic elements. Typically, these have been taken as particles, such as atoms, electrons, protons, quarks, and so on. But you can also add various kinds of fields that extend continuously

2

through space, such as electromagnetic and gravitational. Secondly, these elements are basically external to each other, not only in being separate in space but, more important, in the sense that the fundamental nature of each is independent of that of the other. Therefore the elements don't grow organically as parts of a whole, but rather, as I suggested earlier, they may be compared to parts of a machine. The forms are determined externally to the structure of the machine in which they're working. Now finally, as I also pointed out earlier, the elements interact mechanically, and are therefore related only by influencing each other externally — for example, by forces of interaction that do not deeply affect their inner natures.

In contrast, in an organism, the very nature of any part may be profoundly affected by changes of activity in other parts, and by the general state of the whole, and so the parts are basically internally related to each other as well as to the whole. Of course in a mechanistic view the existence of organism is admitted since it is obvious. But it is assumed, in the way I just described, that ultimately you can reduce it all to molecules such as DNA and proteins, and so on. So eventually the organism is a convenient way of talking about a lot of molecules. They may even say that some new properties and qualities have emerged, but they are always implicit in the molecules. In addition, it's admitted that this goal of a complete mechanistic description is yet to be fully achieved, as there is much that is still unknown. So it's essential for the mechanistic-reductionist program to assume that there is nothing that cannot eventually be treated in this way.

Of course, there is no way to prove this assumption. So to suppose that this assumption holds without limit is bascially an article of faith which permeates the motivation of most of modern science and gives energy to the scientific enterprise. This is a modern counterpart of earlier faith in religious belief based on more organismic types of view, which also in their time gave energy to vast social enterprises. That is, we have not lost the age of faith; we have really changed from one faith to another. And faith is, according to Teilhard de

3

Chardin, just holding the intelligence to a certain world view — that's his definition of faith.

Now how far can this modern faith in mechanism be justified? Of course, there is no question that it works in a very important domain. It has brought about a revolution in our mode of life. Indeed, during the nineteenth century, as I said, there seemed to be little reason to doubt this faith, because of what appeared to be several centuries of successful application leading to vast vistas in the future. Therefore it's hardly surprising that physicists of the time commonly had an unshakable confidence in the correctness of this whole thing. And I may illustrate this by referring to Lord Kelvin, one of the leading theoretical physicists of the time, who expressed the opinion that physics was more-or-less complete in its development. He therefore advised young people not to go into the field, because further work in it would only be a matter of refinements in the next decimal points.

He did however mention two small clouds on the horizon. These were the negative results of the Michaelson-Morley experiment, and the difficulty in understanding black-body radiation. Now we have to admit that Lord Kelvin was at least able to choose his clouds properly, because these were precisely the points of departure for the radical revolution in physics brought about by relativity and quantum mechanics, which overturned this whole conceptual structure. Now this clearly illustrates the danger of complacency about our world views, and makes it evident how necessary it is to constantly have a provisional, inquiring attitude toward them. That is, in some sense, we have to have enough faith in our world-view to work from it, but not that much faith that we think it's the final answer, right?

I couldn't here go into a detailed explanation of how all this took place — this change in view — but I'll give now, beginning with relativity, a brief, non-technical sketch.

I can start by saying that relativity introduced a number of fundamentally new concepts regarding space, time and matter, which are quite subtle. The main point for our purposes here is that the notion of separate and independent particles as basic constituents of the universe had to be given

4

up. The basic notion instead was the idea of the field that spread continuously through space. Out of this you had to construct the notion of a particle. I could illustrate these ideas in terms of the analogy of a flow of fluid such as a vortex. Now within this fluid there is a recurrent, stable pattern. You may abstract it in your mind as a vortex, though there is no vortex. There is nothing but a flowing pattern of water. But a vortex is a convenient word to describe that pattern.

Now if you take two vortices close together, they modify each other producing a different pattern, and eventually, if you bring them together, they merge into one vortex. So you can see, there is an inherent interaction of these patterns, but the basic reality is unbroken wholeness in flowing movement. Separate entities such as vortices, are relatively constant and independently behaving forms abstracted by the mind from the whole in perception and in thought.

This was of course, well known to nineteenth century physicists, but it was generally thought that real fluids such as water were constituted of myriad elementary particles which flowed only in an approximately continuous way, like grains of sand in the hour-glass. The reality underlying the microscopically observed fluid was considered to be a structure of discrete, mechanical elements in the form of particles. But on the basis of the theory of relativity Einstein gave arguments showing that such elementary particles would not be consistent with the laws of physics as developed in his theory. So instead, he proposed a set of continous fields pervading all space, in which particles would be treated as relatively stable and independent structures in limited regions in which the field was strong. Therefore each particle is explained as an abstraction of a relatively independent and stable form, as with the vortex, spread out through space with no breaks anywhere. The universe is seen as unbroken wholeness in flowing movement.

This approach contradicted in an important way the assumption of separate, elementary particles as constituents of the universe, that had been characteristic of the mechanistic view. But still, this theory retained some of the

5

essential features of mechanism, because the fields at different points were regarded as separately existent, and not internally related in their basic nature, and not related to the whole. It was still not anything like the organismic view. The assumption was that these fields are connected only locally — only by infinitesimal steps. The over-all field was viewed as a type of mechanical system that was more subtle than a set of particles, but the field approach was still an important step away from the mechanistic world view, even though it still remained within its general framework.

The quantum theory, however, actually overturned mechanism in a much more thorough way than the theory of relativity. I'll give here its three main features. First of all, all action was in the form of what is called discrete quanta. For example, one found that the orbits of electrons around the nucleus would have to be discrete, as there were no allowed orbits in between, and yet, somehow, the electron jumped from one to the other without passing in between — according to this view. And the light shown on these things was also shown in the form of quanta, and in fact, every form of connection of energy was in the form of quanta. Therefore you could think of it as an interconnecting network of quanta weaving the whole universe into one, because these quanta were indivisible. So this led to some sort of indivisibility of the universe — though it doesn't show up in the large scale because the quanta are very small and, once again, it looks continuous, like the grains of sand in the hour-glass.

Secondly, all matter and energy were found to have what appears to be a dual nature, in the sense that they can behave either like a particle or like a field — or a wave — according to how they are treated by the experiment. The fact that everything can show either a wave-like or a particle-like character according to the context of the environment which is, in this case, the observing apparatus, is clearly not compatible with mechanism, because in mechanism the nature of each thing should be quite independent of its context. But it is quite like an organism, because organisms are very dependent on their context.

6

The third point is that one finds a peculiar new property which I call non-locality of connection. In other words, the connection can be between particles at considerable distances in some cases. This violates the classical requirement of locality — that only things very close to each other can influence one another.

There is another point we can bring out in this connection, which is that the state of the whole may actually organize the parts, not merely through the strong connection of very distant elements, but also because the state of the whole is such that it organizes the parts. It has a certain reality which is indifferent to exactly where the parts are. These are some new features. And all of this shows up in understanding chemistry for example. So when the chemists are using their laws, what underlies them is this peculiar quantum mechanical feature.

Now I want to show how this contradicts the basic mechanistic assumption. Firstly, the action is through indivisible quanta, so as I said, everything is woven together in indivisble links. The universe is one whole, as it were, and is in some sense unbroken. Of course, only under very refined observation does this show up. Now the second point was the wave-particle nature, and the third was non-locality. So, 'ou can see that all these things deny mechanism.

The people who founded quantum mechanics, such as ʃenroedinger, Dirac and Pauli, and so on, all understood this; but since that time this understanding has gradually faded out as people have more and more concentrated on using quantum mechanics as a system of calculation for experimental results, and each time a new text book is written, some of the philosophical meaning gets lost. So we now have a situation where I don't think the majority of physicists realizes how radical the implications of quantum mechanics are.

Now quantum mechanics also says that we don't have complete determinism. That is, the laws are only determined statistically. You cannot tell exactly what is going to happen from these laws. Now this is important too, but perhaps it's less radical than some of these other things, because even from a classical point of view, you can think of laws that are not completely deterministic, such as what's called Brownian motion. So the lack of complete determinism is less radical a change than these others that I've mentioned.

Now how do quantum mechanics and relativity bear on each other? The first point is that the basic physical concepts are quite contradictory. Relativity requires strict continuity, strict determinism, and strict locality. In quantum mechanics, you have to say the opposite — discontinuity, non-determinism, and non-locality. The physical concepts of these two theories have not been brought together, although people are working out equations and methods of doing it mathematically. But the physical meaning has never been made clear.

If you want to look at relativity and quantum theory as being together coherently, we may ask a new kind of question. Instead of focusing on how the theories differ, let's ask what they have in common. What is common to both is unbroken wholeness of the universe. Each has this wholeness in a different way, yet if wholeness is their common factor, that's perhaps the best place to start.

We've seen that each world view holds within itself its own basic notions of order. So we're led naturally to the question: Is it possible to develop a new order that is suitable for thinking about the basic nature of the universe of unbroken wholeness? This would perhaps be as different from the order of mechanism as the latter is from the ancient Greek order of increasing perfection. Now we won't necessarily return to ancient Greek or organismic theories, but to something new, perhaps different from both.

This brings us to the further question though: What is order? Now we do presuppose that there is some kind of order — so a general and explicit definition of order is not actually possible. You see, to begin you must already understand something about order, because just to talk, you must have some understanding of what order is and what meaning is. You can take a few examples to illustrate this — the order of numbers, 1,2,3,4; the order of points on a line; the order of functioning of a machine; the subtle order of functioning of an organism; the many orders of tones in music; of time; the order of a language; the order of thinking, and so on. You see, there are all sorts of orders that are more and more subtle. The notion of order covers a vast and unspecifiable range. So I'll take it for granted that we already know, tacitly, something of the notion of order. And then our whole point is to bring it out.

Most of this tacit notion of order is based upon perceptual experience, as you see from the examples. One could ask if there is not an analogy in our experience that would discuss the order of unbroken wholeness. Here I could point out that the operation of scientific instruments has often played a key part in helping to make certain notions of order clear. The lens, for example is a device that makes an image.

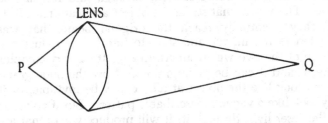

Point P is imaged by the lens into point Q, roughly — it's not exact. Now, in this way you can consider together all the image points Q, and you'll have a photograph of the object. This constitutes a kind of knowledge of the object in which we are stressing the point-to-point correspondence between the image and the object. Therefore, you are stressing the concept of points. With the aid of telescopes, microscopes, very fast or slow cameras, and so on, this kind of knowledge through correspondence of points, could be extended to things that are too far away, too small, too fast, too slow, and so on, to be seen with the naked eye. Eventually in this way you would be led to think that everything could ultimately be known in the form of separate elements. This shows that instruments based on the lens have given a great impetus to the mechanistic way of thinking, not only in science, but in every phase of life.

The implicate order: a new approach to reality

I could ask: Have any instruments been developed that would similarly help point vividly to the way of thinking that is compatible with unbroken wholeness? Now it turns out that there are several. I'll begin by describing the holograph which was invented by Dennis Gabor. This name is based on two Greek words — *holo* meaning whole, and *graph* meaning to write. The holograph writes the whole. From this point of view a lens could be called a 'merograph', which writes the parts, and a telegraph, I suppose, writes far across. This instrument depends on another device called a laser, which produces a beam of light in which the waves of light are highly ordered and regular, in contrast to those of ordinary light where they are rather chaotic. Light from a laser falls on a half-silvered mirror. Part of the waves reflect and part of them come straight through and fall on the object. The waves that strike the object are scattered off it, and they eventually reach the original beam that was reflected in the mirror and start to interfere, producing a pattern of the two waves superimposed. It's a very complex pattern, and it can be photographed. Now the photograph doesn't look like the object at all. It may be invisible, or it may look like a vague indescribable pattern. But if you send similar laser light through it, it will produce waves that are similar to the waves that were coming off the object, and if you place your eye in the right spot you will get an image of the object which will apparently be behind the holograph, and be three dimensional. You can move around and see it from different angles, as if through a window the size of the beam.

The point is that each part of the holograph is an image of the whole object. It is a kind of knowledge which is not a point-to-point correspondence, but a different kind. By the way, if you use only a part of the holograph, you'll still get an image of the whole object, but you'll get a less detailed image, and you'll see it from a more limited set of angles. The more of the holograph you use, the more of the object you can see, and the more accurately you can see it. Therefore every part contains information about the whole object. In this new form of knowledge information about the whole is enfolded in each part of the image. I can give an idea of enfoldment in a preliminary way by thinking of taking a sheet of paper, folding it up many times and, say, sticking pins in it, cutting it, and unfolding it, and you've made a whole pattern. So the pattern is enfolded, then it unfolds. In some sense the holograph does that.

Of course, in this example the photograph is only a static record of the light, which is a movement of waves. The actuality that is directly recorded is the movement itself in which information about the whole object is dynamically enfolded in each part of space, while this information is then unfolded in the image. A similar principle of enfoldment and unfoldment can be seen to run through a wide range of experience. For example, the light from all parts of the room contains information about the whole room and, in a way, enfolds it in this tiny region going through the pupil of your eye, and it is unfolded by the lens, and the nervous system — the brain — and somehow, consciousness produces a sense of the whole room unfolded in a way which we don't really understand. But the entire room is enfolded in each part. This is crucial, because otherwise we wouldn't be able to understand what the room was — the fact is that there is a whole room, and we see the whole room from each part. The light entering a telescope, similarly enfolds information about the whole universe of space and time. And more generally, movements of waves of all sorts enfold the whole in each part of the universe.

This principle of enfoldment and unfoldment may be observed in a more familiar context. For example, information out of which a television image is formed is enfolded in a radio wave which carries it as a signal. The function of the televison set is just to unfold this information and display it on the screen. The word 'display' also means to unfold, but for the purpose of showing something, rather than for its own sake. This is especially clear in the older televison sets that had an adjustment for synchronism, so when they went out of adjustment you could see the image folding up, and as you readjusted it, it unfolded.

In the mechanistic world view all these examples are well

known, but they are explained by saying that the primary reality is ultimately the basic set of independently existent elements — particles and fields — while the enfoldment and unfoldment is only a secondary aspect. They say it's not very important. What I'm suggesting here is that the movement of enfolding and unfolding is ultimately the primary reality, and that the objects, entities, forms, and so on, which appear in this movement are secondary.

Now how is this possible? As I've already pointed out, quantum theory shows that the so-called particles constituting matter are also waves similar to those of light. One can, in principle, make holographs using beams of electrons, protons, and so on, as well as sound waves — which has been done. The key point is that the mathematical laws of the quantum theory that apply to these waves, and therefore to all matter, can be seen to describe just such a movement in which there is a continual enfoldment of the whole into each region, along with the unfoldment of each region into the whole again. Although this may take many particular forms — some known and others not yet known — this movement is universal as far as we know. I'll call this universal movement of enfoldment and unfoldment 'the holomovement'.

The proposal is that the holomovement is the basic reality, at least as far as we are able to go, and that all entities, objects, forms, as ordinarily seen, are relatively stable, independent and autonomous features of the holomovement, much as the vortex is such a feature of the flowing movement of a fluid. The basic order of this movement is therefore enfoldment and unfoldment. So we're looking at the universe in terms of a new order, which I'll call the enfolded order, or the implicate order.

The word 'implicate' means to enfold — in Latin, to fold inward. In the implicate order, everything is folded into everything. But it's important to note here that the whole universe is in principle enfolded into each part actively through the holomovment as well as all the parts. Now this means that the dynamic activity — internal and external — which is fundamental to what each part is, is based on its enfoldment of all the rest, including the whole universe. But of course, each part may unfold others in different degrees and ways. That is, they are not all enfolded equally in each part. But the basic principle of enfoldment in the whole is not thereby denied.

Therefore enfoldment is not merely superficial or passive but, I emphasize again, that each part is in a fundamental sense internally related in its basic activities to the whole and to all the other parts. The mechanistic idea of external relation as fundamental is therefore denied. Of course, such relationships are still considered to be real but of secondary significance. That is, we can get approximations to a mechanistic behaviour out of this. That is to say, the order of the world as a structure of things that are basically external to each other comes out as secondary and emerges from the deeper implicate order. The order of elements external to each other would then be called the unfolded order, or the explicate order.

The usual way of looking at things is, therefore, turned upside-down, and that's how we arrive at the notion of the implicate order. The holograph is, of course, only a particular example of an implicate order. Its value in the present context is that it provides a good analogy as to how the implicate order is relevant to the quantum behaviour of matter. The analogy is particularly good because, as I've said, the laws of the propagation of the kinds of waves that are associated with basic quantum laws are also capable of being compatible with the theory of relativity, and therefore we see that the implicate order is able to have a significant bearing on both of the two most fundamental theories of modern physics.

But of course, analogies are necessarily limited, since by their very nature they are similar only in some ways to what they are representing and are different in other ways. One of the principal limits of the analogy of the holograph, at least as it's usually analysed, is that it does not adequately take into account all of the quantum properties of the waves that are involved. In particular, what it fails to consider is that the energy of these waves is in discrete units, or quanta, called photons. Now usually there are so many of them that this is not important. But if we wanted to be very accurate, this would be important. The holographic analogy still misses some of the essential features of quantum mechanics. To make an accurate analogy one would have to also use modern, relativistic quantum theory, and this would lead to questions that are much too abstract and complex to be treated here. But the point about analogies is that they are always limited, and if they were not limited, they would not be distinguishable from the thing itself. So we can keep on using analogies which are almost like metaphors to help get across what is meant.

Now as another analogy, I think you've all seen computer games. You have a television screen which you could call an implicate order because, as I've just explained, out of this can be unfolded all sorts of forms according to what goes in. But if this screen is connected to a computer, then the computer will unfold forms, for example, spaceships and so on, according to its program, and you can see now that the computer enfolds the information needed to determine the spaceships. So there are two implicate orders — one, the implicate order of the screen, and two, the way in which the information is enfolded in the computer. Thirdly, there are the buttons that the player presses, and then we have the person who plays it — that's the third implicate order. He enfolds further, and he of course, is affected by what's on the screen, and so it goes around. So the three together make a kind of unit. And it becomes so absorbing that in some cases they really are a unit. Now this is a good analogy as to how the quantum mechanical field theory works, because the first implicate order is like the field, and there is a super-implicate order which organizes the field into discrete units which are particle-like. Without that super-implicate order however, the field would just spread out without showing any particle-like qualities.

It's possible to produce an indefinite number of additional analogies, but what I want to do instead is to discuss the more general significance of the implicate order beyond physics. What I want to say is that if you look beyond physics you will find that orders similar to this implicate order are really quite common in experience. In fact, this idea of enfoldment is an ancient idea. It was known in the East a long time ago.

If you take the example of a living being such as a plant grown from a seed, the seed makes a very small contribution to the substance of the fully grown plant and to the energy needed to make it grow. These come from the air, the water, the soil and the sunlight. According to modern ideas of genetics, the seed has information, if you like, in the form of DNA which is transmitted to the matter out of which the plant is eventually formed. Now we have already been led to use the notion of the implicate order for matter in general. We see how it is constantly enfolding again into the background. You may think of an electron as unfolding from this background at a particular position, then it folds back in again, and another unfolds nearby, and it enfolds again, and another one, and another one, and gradually it looks like a track of a single electron. You can see the discontinuity here because the places of unfoldment need not be continuous. And you can understand why there can be discontinuity and also continuity — wave-like qualities — coming from the unfoldment. So we see that inanimate matter is constantly recreating itself through enfoldment and unfoldment — replicating itself, if you will — in the form of inanimate matter. That's the proposal. Now with the further information from the seed, it unfolds to make a plant instead, which can then make seeds for new plants. You can look at it as a continuous process of unfoldment that can be modified by new orders coming from the genetic structure, so that it will unfold into a considerably different being.

Let's go on to discuss consciousness, which we take to include thought, feeling, desire, will, impulse to act and an unspecified set of further features, such as awareness, some of which we may discuss. The question is: Do we find an implicate order in consciousness? To answer this question I will first consider the process of thought. In describing this process we may refer to thoughts that are implicit. The word 'implicit' has the same root as implicate, and this suggests that a given thought may somehow contain other thoughts that it implies — that is, that it enfolds. Such implication may be, in some cases, equivalent to entailment or inference if it obeys the rules of logic. But this is only a special case of implication, like that of a regular track. There may be implications which produce very regular tracks, or more irregular tracks, so that there could be leaps in thought, and so on. So implication has a much wider range of meanings, going from mere association to a sense that one thing goes with another, and to a tacit, or unstated, ground of reason supporting the thought that is implied. All of these may be regarded as enfolded within the thought in question and are capable of emerging from it through unfoldment.

Here I could add that language, which is essential to the communication of thought and to its precise determination, may also be seen as an implicate order. After all, the word is only a sign or a symbol of very little significance in itself. What is more important is its meaning. Generally this is determined only by a much larger over-all context. For example, the meaning of a given word may be affected by other sets of words, not only near to it but even quite far away, and this suggests that the meaning of each word, and indeed each combination of words, such as a sentence or a paragraph, is ultimately unfolded into the whole content that is communicated. Such a notion is suggested even more strongly by the fact that often one can sense that the whole sequence of words seems to flow out of single momentary intention without the need for conscious choice for their order, essentially as if they had unfolded from something that was already there in the intention.

As a further interesting example, there is the fact that without the need for a search in memory we can sense whether a word is in common usage in the language or not. Thus nouns formed out of verbs, such as 'alternation', generally have in common usage verbs that correspond with them, such as to alternate. We know immediately though, that in certain cases they do not. For example, 'alteration' does not have the corresponding form, 'to alterate'. You don't have to search to find that out. So it suggests that some features of this language are, as it were, enfolded in the whole, although that doesn't necessarily explain all of them. The immediate availability of this knowledge, then, suggests that you can think of the totality of a given language as an undivided whole from which the various words and their potential meanings all unfold. A reasonable case can therefore be made for the proposal that thought and language form an implicate order. But these also enfold feelings and, vice versa, feelings enfold thought. Language, you see, is implicit in feelings and thoughts and words. The thought of danger unfolds into a feeling of fear, which unfolds into words communicating the feeling, and to further thoughts, and you see all of this mutual enfoldment.

Thoughts and feelings also enfold intentions. These are sharpened up into a determinate will and the urge to do something. Intention, will and urge unfold into more action, which will include more thought if necessary. So all the aspects of the mind show themselves as enfolding each other, and transforming into each other through enfoldment and unfoldment. And therefore we have a view in which the mind is not regarded as broken up dualistically or multiply into independently existent functions or elements like thought and feeling, because in enfoldment each aspect is internally related to the other rather than externally.

If you're attentive you can see quite a few other things that indicate this enfoldment. I would like to suggest we consider listening to music. Your attention shows that while any given note is being played several preceding notes are still present in awareness as a kind of immediate after-echo, or reverberation. This is to be distinguished from memory, which is recalled or re-collected from a more permanent repository. Remembering notes a minute apart is not perceived as music, and most of the music is then lost. The notes must somehow be present together. One can sense that each note, as it starts to fade and turn into a diminishing sequence of after-echos, is in some way enfolding into various aspects of consciousness including emotions, associations of various kinds, impulses to move, and so on. I'm suggesting here that this may be seen as a kind of enfolded order. That is to say, one can sense the co-presence of after-echoes and other derivatives of several notes in different degrees of enfoldment. This is similar to the structure of the enfoldment of many waves into one in a holograph. The essential point is that the simultaneous co-presence of several notes, and possibly in some sense even some distant ones, has its origin in the sense of flowing movement of the theme, along with the preservation of its essential identity, which explains why notes that follow each other only after long intervals generally convey neither a sense of flowing movement nor a preservation of identity.

Now there is another example, brought up by Michael Polanyi, of bicycle riding. In order to remain stably upright, one must turn into the direction in which one is falling. Polanyi has pointed out that a simple calculation based on the laws of physics shows that, if the bicycle is being ridden properly, its angle of tilt and the angle at which the wheel is turned are related by a certain formula. But of course, any attempt to follow this formula would get in the way of actually riding the bicycle. What is of key significance is that the over-all movement that results and brings about approximately following the formula is the outcome of an entirely different level of activity involving muscles, nerves and brain. It is extremely complex and subtle, and evidently you cannot describe it in any explicit way. Polanyi called this 'tacit knowing', rather than explicit knowing. I would like to propose that this may be regarded as a kind of implicate order which unfolds into an explicate order of the motion of the bicycle as described by a formula. The law of the explicate order therefore emerges as an abstraction of what is actually a certain feature of a much larger implicate order.

Evidently this kind of tacit knowledge is very important in every phase of life. In fact, without tacit knowledge ordinary knowledge would have no meaning. In fact, when we talk, most of the meaning is implicit or tacit. And also the action which flows from it is implicit or tacit. In fact, even to talk or to think — although thinking may be explicit as it forms images — the actual activity of thinking is tacit. You cannot say how you do it. If you want to walk across the room, you cannot say how it comes about, right? It unfolds tacitly.

On the basis of all of this I would then propose for further discussion the notion that both mind and matter are ultimately in implicate orders, and that in all cases explicate orders emerge as relatively autonomous, distinct and independent objects, entities and forms, which unfold from the implicate orders. This means that the way is opened up for a world view in which mind and matter may consistently be related without adopting a reductionist position.

Here we are going to say that mind and matter both have reality, or perhaps that they both arise from some greater common ground, or perhaps they are not really different. Perhaps they interweave. The main point, though, is: because they have the implicate order in common it is possible to háve a rationally comprehensible relationship between them. In this way we can leave open the possibility of acknowledging the differences that may be found between the mental and material sides without falling into dualism.

This question of how mind and matter are related has long been one that has perplexed those who have seriously enquired into it. Descartes gave an especially clear and sharp formulation of the difficulties. He considered matter as extended substance — that is, existing spread out in space in the form of separate objects. Mind he discussed in terms of thinking substance which is not separate and extended — that is, thoughts of distinct objects are not themselves spread out. You see, we can make clear and distinct thoughts, yet they don't exist as separate and extended elements in any kind of space.

Descartes felt that the two substances were so different that there was no way to formulate their relationship clearly. The problem of how they are related was to be solved by bringing in God who created both, and who is thus the ground of their connection — that is, God puts clear and distinct thoughts into our minds which may correspond correctly to the separate objects of space. He also thought that maybe the pineal gland would connect mind and matter, but that's not very consistent because he only puts the problem into the pineal gland and doesn't say how it can do that — connect such different things.

Since the time of Descartes the idea that problems of this kind can be solved by an appeal to the action of God has been dropped. But it has not generally been noticed by those who go on with Cartesian mind-matter duality that this leaves the whole problem of how the two are related unsolved. Or perhaps it has been noticed, but it has been more or less put aside.

The implicate order suggests a possible solution of this Cartesian duality which has pervaded much of human thinking over the ages. Instead of saying that there are two orders — the explicate order of extended structure, and something like an implicate order of thinking — we are proposing, to a large extent, on the basis of an understanding of recent developments in physics, that matter also is that way. And if we were to extend it to say that brain matter and nerve matter are that way, then in some way perhaps, mind and matter interweave. And perhaps something analogous to mind might exist in inanimate matter, at least implicitly, just as life is implicit in inanimate matter. Given a seed it forms animate matter instead.

And somehow mind is implicit in inanimate matter. Given the proper conditions it unfolds and forms living beings who might even be conscious. And that might suggest — this is something we'll go into — that the mental and the material are two sides of one reality.

The division between mind and matter, or the observer and the observed, has produced very serious consequences in attempting to see that the world is a whole, because even if you are thinking of wholeness, you are thinking of an observer who is looking at this wholeness, and this creates a division. This starts to break up the whole, because you identify with one part of it, and then there is another part you are not identified with, and therefore the whole is broken up in two. And then it breaks up further, because there are many observers, and each observer is an external object for all the others. The many parts obtained in this way are related, and you have to break things up even more in order to understand their relationships. So the implicate order can be important as a way of seeing how this particular problem might be dealt with.

But let me emphasize that to have an approach of wholeness doesn't mean that we are going to be able to capture the whole of existence within our concepts and knowledge. Rather it means firstly that we understand this totality as an unbroken and seamless whole in which relatively autonomous objects and forms emerge. And secondly it means that, in so far as wholeness is comprehended with the aid of the implicate order, the relationships between the various parts or sub-wholes are ultimately internal. This notion is suggested also by an organismic point-of-view; but as I've said, there is no way to exclude the possibility that organisms have a mechanistic base in their supposed constituent particles. But if we say that the particles themselves haven't got a mechanistic basis, then why should the organisms have it? It would be peculiar to say that the particles of physics are not mechanistic, but as soon as they make organisms they are mechanistic.

It is important to keep in mind here that the whole and its parts are correlative categories — that each implies the other. Something can be a part only if there is a whole of which it can be a part. To understand this correlation of whole and parts, I want to return to the notion of the holomovement. Within the holomovement, as I've said, each part emerges as being a relatively independent, autonomous and stable sub-whole, and it does so by virtue of the particular way in which it actively enfolds the whole and therefore all the other parts. Its fundamental qualities and activities both internal and external are essential to what it is and are thus understood as determined basically in an internal relation, rather than in isolation and external relation.

This internal relationship is most directly experienced in consciousness. The content of consciousness of each human being is, evidently, an enfoldment of the totality of existence, physical and mental, internal and external. This enfoldment is active in the sense that it enters in a fundamental way into the activities that are essential to what a human being is.

21

According to the content of his consciousness he acts, whether it's right or wrong. Each human being is therefore related to the totality, including nature and the whole of mankind. He is also therefore internally related to other human beings. How close that relation is, has to be explored. What I am further saying is that the quantum theory implies that ultimately the relationship of parts and whole — of matter in general — is understood in a similar way.

And perhaps I should also add here that in each sub-whole there is a certain quality that does not come from the parts, but helps organize the parts. So the implicate order does not deny the significance of parts or sub-wholes, but rather it treats them in its own way as relatively stable, independent and autonomous. Wholeness is seen as primary while the parts are secondary in the sense that what they are and what they do can be understood only in the light of the whole.

I could summarize this in the principle: The wholeness of the whole and the parts.

And the opposite principle: The partiality of the parts and the whole.

Both principles have their place. But I will make an assertion: The need to accentuate the wholeness of the whole and the parts.

This assertion is needed, because we have to be careful not to assert wholeness too strongly, or else we will just simply create opposition to something that is perfectly valid, namely mechanism in a limited area. The difference is not whether the parts are included, but what is given primary emphasis. This is rather as in a musical composition, where the entire meaning depends upon which theme has a major or dominant role, and which is minor or secondary. This is a basic feature of communication at the metaphsyical level. To some extent it's an art form. You cannot get a precise communication, but it is implicit or tacit, what is being communicated. And therefore the form in which it is put is crucial. The form must be appropriate to the content.

There is a danger in seeing mechanism as totally destructive and saying that we must only discuss the whole. For that also is a partial view and, in fact, it is almost another form of mechanism. So we are just asking: Where do we put the ultimate emphasis? But of course, if you don't want to do metaphysics, which is a view of the nature of reality as a whole, then you don't have to accentuate either principle. You'll say, you're just going to take these two principles as practical principles to apply wherever you think they're appropriate. Then they become maxims, which may apply here or there. You choose your maxim according to where it works. However, we'll see as we go along that this attitude cannot be maintained indefinitely, and that ultimately we must regard one of these two principles as the major theme and the other as the minor theme.

This approach of wholeness could help to end the far-reaching and pervasive fragmentation that arises out of the mechanistic world view. One can obtain a further understanding of the nature of such fragmentation by asking, what is the difference in the meaning of the word 'part' and 'fragment'. A part, as I said — whether mechanical or organic — is intrinsically related to a whole, but this is not so for a fragment. As the Latin root of the word indicates, and as the related English word 'fragile' shows, to fragment is to break up or smash. To hit a watch with a hammer would not produce parts, but fragments that are separated in ways that are not significantly related to the structure of the watch. If you cut up the carcass of an animal as in a butcher shop, this produces not parts of the animal but fragments again. So what I'm trying to say is that we have a way of thinking that produces irrelevant breaks and fragments, rather than seeing the proper parts in relation to the whole.

Of course there are areas where it is appropriate to produce fragments. If you can crush stones in order to make concrete, that's perfectly alright. There are things that should be broken down into fragments. But what I'm discussing here, quite generally, is an inappropriate kind of fragmentation that arises when we regard the parts appearing in our thought as primary and independently existent constituents of all reality including ourselves — that is, that corresponding to our thoughts there is something in reality. Then a world view such as mechanism, in which the whole of existence is considered as made up of such elementary parts, will give strong support to this fragmentary way of thinking. And this in turn expresses itself in further thought that sustains and develops such a world view. As a result of this general approach, man ultimately ceases to give the divisions the significance of merely convenient ways of thinking, indicating relative independence or autonomy of things, and instead he begins to see and experience himself as made up of nothing but separately and independently existing components.

Being guided by this view, man then acts in such a way as to try and break himself and the world up so that all seems to correspond to this way of thinking. He therefore obtains an apparent proof of his fragmentary self-world-view, but he doesn't notice that it is he himself, acting according to his mode of thought, who has brought about the fragmentation which now seems to have an autonomous existence independent of his will and desire.

Fragmentation is therefore an attitude of mind which disposes the mind to regard divisions between things as absolute and final, rather than as ways of thinking that have only a relative and limited range of usefulness and validity. It leads therefore to a general tendency to break up things in an irrelevant and inappropriate way according to how we think. And so it is evidently and inherently destructive. For example, though all parts of mankind are fundamentally interdependent and interrelated, the primary and overriding kind of significance given to the distinctions between people, family, profession, nation, race, religion, ideology, and so on, is preventing human beings from working together for the common good, or even for survival.

When man thinks of himself in this fragmentary way, he will inevitably tend to see himself first — his own person, his own group — he can't seriously think of himself as internally related to the whole of mankind and therefore to all other people. Even if he does try to put mankind first, he will perhaps think of nature as something different to be exploited to satisfy whatever desires he may have at the moment. Similarly he will think body and mind are independent actualities, thought and feeling, and so on, and he begins to think to divide these up, each to be treated separately. Physically this is not conducive to over-all health, which means wholeness, and mentally, not to sanity which also has a similar meaning. This is shown, I think, by this ever growing tendency to break up the psyche in neuroses, psychosis, and so on.

Well, to sum up, fragmentary thinking is giving rise to a reality that is constantly breaking up into disorderly, disharmonious and destructive partial activities. It therefore seems reasonable to explore the suggestion that a mode of thinking that starts from the most encompassing possible whole and goes down to the parts as sub-wholes in a way appropriate to the actual nature of things, would help to bring about a different reality, one that was more harmonious and orderly and creative. And in this discussion here, I have tried to show that physics provides some justification for doing this. And in fact it is more justified than the mechanistic view if you go into physics deeply. But of course before it really changes things — to think differently — this thought must enter deeply into our intentions, actions, and so on — our whole being. That is, we will actually have to mean what we are saying. To bring this about requires an action going beyond what we have just discussed. The main point then is that your world views — it's really a self-world-view because it includes yourself — have a tremendous effect on you. Even people who don't think they have self-world-views have them tacitly. And the general prevalence of mechanism has helped to give rise to fragmentation. The fact is however that even when people held an organic point of view in ancient Greece they also fragmented, so there's more to it than that. The self-world-view has to be pursued carefully into the whole question of the division of mind and matter to see how fragmentation comes about. Such fragmentation doesn't come only from philosophical views, but philosophical views can either contribute to it or contribute the other way. But of course, to understand this whole question, much more is required.

Question: Have you heard of the work of Mary Douglas the social anthropologist? She has done a lot of work about how hard it is for us to get out of our own categories. She claims that whenever we transcend the categories, it unleashes pollution. A kind of an alarm, as when you break down the system itself. She claims that whenever we make these forms that constitute our classifications they are quick with energy, and whenever we try to go against them something terrible will ensue.

Professor Bohm: Yes, well that could be so. You see, the forms that we have in our world view are charged with tremendous energy, and presumably when that world view is challenged, say by a scientific world view or a religious world view, a tremendous explosion may take place and people fight over it to the death, right? Nevertheless it may be necessary to challenge these world views if they are wrong. There is a risk in doing so, but there is also perhaps a greater risk in not doing so, because if we go on with a fixed, rigid world view it will lead us to the edge of the abyss, right? As we are now approaching.

Q: My impression is that to a great extent what you are talking about is a pervasive confusion — that we mix up the fragments and think that they are wholes, then behave as if they are wholes.

Bohm: Yes, well that's because we are thinking that the parts in our thought each corresponds to a sub-whole; but more deeply, because we are taking this thought as an exact representation of reality, we are imposing it on reality where in general it won't hold. So in attempting to impose this thought on reality rigidly, we start to try to break reality up.

Now if I thought that this (piece of chalk) was made of two parts and I moved it, of course the parts would move together; but if I continued to insist that it's made of two parts I'd have to break it, so it would then be two parts.

Now you see, if we say there are two nations, that's the
same kind of problem. You see, the people in the two
nations may not be very different, like France and
Germany, right? Nevertheless they insist they are
absolutely different. One says, *'Deutschland über Alles'*,
and the other says, *'Vive la France'*, and then they say,
'We must establish rigid boundaries; we must set up
tremendous big fences across these boundaries; we must
destroy anything to protect them,' and you had the First
World War. Of course, each part had its commercial
interests, and so on, despite the fact that they were
interdependent, and probably they would have prospered
much more if they had allowed free flow, as for example
happened with the states of the United States. So if you
think that there are two parts, then you will impose it.
Though if you cross the boundary no division is visible; the
people are not very different, and if it had happened
historically, by accident, that the two had been one, there
would have been no such thing.

Q: Could the whole be one without being broken into
parts?

Bohm: That's not broken. You see, I'm trying to say that
the whole divides into parts, and they are natural. The
whole and the parts are correlative categories — the parts
are sub-wholes. Now there's a difference between breaking
and natural division. The cells may divide naturally, but if
they are smashed it's quite different. The attempt to
impose a line of thought too hard will tend to lead to
arbitrary division. Even the parts are not absolutely
divided from each other, because you can see that
underneath they come out of a common whole —
unfolding — but only relatively so. If you think of a table,
it looks quite divided from the people. It's made of atoms;
but actually there's no place where the table ends. If you
try to think of it at the atomic scale it would look very
vague, and it goes into the air and into the people, and
also, if you were to think of the quantum mechanical

27

nature, all would be unfolding. We could say that this notion of division into parts is an abstraction which we can apply up to a certain limit. Thus for example, we divide land into fields for different purposes, and we call them different parts. That is convenient and useful and correct up to a point; but if we take it too seriously it's wrong, because every part depends on every other part. You see, the enfolded relation of the whole eventually has to prevail.

Now the same thing holds, let's say, in ecology. You divide the world into parts, but you find the division doesn't hold — that pollution occurs in one place and goes to another, and problems created in one place spread out, and little things happening here and there all add up all over. Therefore this idea of dividing things into parts is of limited value. We are not discarding it, but we are saying it's got to be used intelligently. Otherwise, you will start smashing things up — that is fragmentation. I want to make a sharp distinction between a part and a fragment. There are some things which should be smashed up, so I'm not totally against smashing up.

Q: I find my mind tending to want to separate cause and effect. Sometimes it's fair, but probably there's a tendency often to want to find — to see the effect — like to see our world view having an effect on the nature of the world.

Bohm: That's a very subtle question. We have to discuss in what sense is our world view a part of the world, and I think it's probably a bit late for that tonight. (*laughter*) Roughly, I would say that we have to see our world view as an active part of the world. But what is the real relation of parts and whole — or cause and effect — is there a division?

Q: Would the answer to that include something of Rupert Sheldrake's morphogenic ideas?

Bohm: Well, that could be part of the answer; but I think

it's a more general question about the nature of our thought, the nature of how it participates in reality. Is it part of reality? And also the nature of cause and effect, is there a division?

Q: I don't know how long you want to speak, but I got bogged down in the first ten minutes of your talk, and reading the first chapter of your book (*Wholeness and the Implicate Order*), I suddenly found it getting beyond me. What I really can't seem to do is to actually grasp what we are really talking about with this basic theme of enfoldment and unfoldment. Can I just take two examples that you brought up? One is the holograph. Now as far as I know, and I don't really know the details, I believe that if you get a holograph, which is a photograph, it is possible to analyse exactly what it is a holograph of, and if you pass laser beams through it, what it is will actually appear. In other words, you know that there is a simple explanation for that in purely mechanistic terms. So this is the first question: What actually are we gaining by saying, well now, this is an example of unfoldment? Can I just move on to plants? Because when you have a plant growing, I think you have to admit that you can't actually predict in mechanistic terms how exactly that plant is going to grow. There are some arbitrary factors too complicated to pin down. What are we gaining when we say don't explain it in those terms but see it as an unfoldment of something that was previously enfolded? I think I really mean that to talk anything about wholeness, I would like to feel I can actually grasp in my mind what that operation is.

Bohm: Yes. Now with the holograph, you see, it's true you can give a mechanistic explanation. I was merely using the holograph as an analogy of enfoldment to get a picture. You can give a mechanistic explanation if you don't go too deeply, but if you look at the quantum nature of the waves that you use in explaining them, you'll find that there is no mechanistic explanation. It's only when you treat the waves as a classical wave that you have a mechanistic explanation, and I was merely using this as an analogy to help get across what the meaning of enfoldment is. Now the reason for this enfoldment is that the laws of quantum mechanics have no mechanistic explanation finally. And since everything else is supposed to be based on that, it means that we cannot actually give a mechanistic explanation that holds all the way. There only is an approximation. Now if we are trying to see the thing deeply — the nature of what is — we have got to look differently, and I am proposing that if you begin with enfoldment you get a feeling of what is most basic, and you can then explain mechanism as an approximation to unfoldment.

Q: Can unfoldment be seen as predictive?

Bohm: It's not only a question of predicting but understanding. This is an important point. In earlier science the idea was to understand the universe, and to predict it also. Now if you emphasize prediction, we are in the fragmentary point of view again. You are saying that understanding is not important; the important point is to predict. Now that's important for technology and for various purposes, but I'm trying to say that if we adopt that world view all the way, we are going to take a mechanistic attitude to each other and to everything, and we are going to have some difficult consequences.

Q: I'm not exactly wanting to predict, but on the other hand if your prediction comes out correct, that gives you some reassurance that you are on the right track.

Bohm: This holomovement point of view is in principle capable of treating a wider range of problems. For the moment it is merely another way of looking at what is treated by quantum mechanics. Now quantum mechanics enables you to predict certain things — the probability of them — but it does not enable you to understand what it means. That is, it's merely a set of rules like a cook book. If you turn the crank you'll come out with the answer. There's no real way of intuitively understanding what underlies it. So, I'm trying to say that if you try to understand intuitively what underlies quantum mechanics, this image of a holograph will give you a feeling for it.

I think it's important that people understand scientific ideas intuitively for many reasons. One of them is that the only way of conveying it to the general public is intuitive; otherwise we must treat scientists as super witch-doctors who work out all these formulae and produce magical results, and you'll have faith in them. I think it's important that the general public should have some understanding, and in the past they did. The second point is that it's part of your world view that by applying this mechanistic philosophy and saying that the fundamental particles are mechanical when in fact one shows they cannot be, we are affecting our whole way of approaching the world and ourselves. And this has a profound effect on the way science is done, and the way society is organized, and the way people are related. Therefore it is important to get an intuitive notion, which is the notion by which you will be moved. You see, the fact that you can predict certain things has very little ... oh, except say, maybe we can gain something by predicting — but the intuitive feeling that is now gotten across by science is that mechanism is the nature of reality. So that gives the notion that science backs up mechanism. But in fact it's just a philosophical idea and not backed up by the whole of science; the most fundamental features of science don't back it up.

Q: I'll sleep on that!

Bohm: Yes. One more question because its very late.

Q: Something I've been most drawn to when we've been talking about cause and effect is an idea that I came across in a book which I found very exciting, which was the idea that the explicate cause and effect — the explicitly, seemingly mechanistically related events — may refer to an implicate, higher dimensional going-on — some implicate happening, some implicate presence, happening.

Bohm: Yes, well, that will require some explanation, go ahead.

Q: That idea and the intuitive ramifications of that idea I find very exciting. And just to open the thing up — maybe it will develop in the time we spend together — one aspect of that I'd just like to bring up is the possibility of seeing all the different things that people do in, say, transformational consciousness, as perhaps not causally related, but referring to something of a larger happening that maybe we can't have a grasp of, but that we do have, as many parts of the hologram — taking the thing as a whole we get an insight into what's going on.

Bohm: Yes. If you would raise that question tomorrow....*(laughter)*

Peter Garrett: As David Bohm was encouraging there, maybe you could take a sheet of paper and write a specific question or questions that we could then feed back and have a look at tomorrow. Another thing just to add is that you may feel the question you have is a little bit irrelevant to the ideas that are being talked about. We were talking at lunch time about a letter that had been mislaid for two years which turned out to have one thought in it which was very helpful to the development of some ideas that you (David Bohm) were working with — the person in Australia who had a comment which was very useful — so your ideas might seem a little off the side, or simple, or something, but it may be extremely useful. So why don't you write it down anyway, and see what happens. Thank you very much.

DISCUSSING THE IMPLICATE ORDER

Following the first session the organizers were informed that some of the participants had found it very difficult to see or hear Professor Bohm from their seats at the back of the room. Late that night, with the excellent cooperation of the Three Ways Hotel management, a small platform was devised and the lighting and seating rearranged.

Bohm: Well, it's strange to be elevated and illuminated. I received a very big list of questions. Some of them are quite long, and they're all very good questions actually. I appreciate them. I've read them, and I think that it's good that you wrote them. However I think if I were to answer them in that way this would not be a dialogue; I would be teaching you, or I would be trying to give my views one after another. So perhaps it's a good idea for you to write the questions so that I know what's on your mind, but that we shouldn't actually proceed from the written questions. I want to thank you again for all the trouble you took writing these questions.

Now I'll begin by asking: What is inquiry? We want to inquire today. And I don't know the answer to most of your questions. *(laughter)* We are sort of pushing far beyond where I have gone. Perhaps together we can go forward a little.

To begin with, why do we inquire? Does anybody have an idea? *(pause)* Because there is something that we do not understand; something is not working out right. As long as things are working out right there is no reason for a question. Then you raise a question. Where does the question come from? It must come from the same general mind which produced the situation that needed the inquiry — that is, we've been doing something wrong somewhere; our thoughts are wrong or our actions are wrong, and out of that mind we have made a question.

That question will contain presuppositions, and we all know what presuppositions are. I can see this from the level of questions. These presuppositions are largely unaware. By the time we become aware of them we can call them assumptions. We presuppose all sorts of things. You see, when I'm going to walk on the floor I presuppose that it will support me, but it might not. I presuppose that the road is going to be OK, and that the car will work, and so on. Then, as soon as you find it's not so, you must look again, and begin to see where your presuppositions are wrong, and change them. So in an inquiry we are looking into our presuppositions.

Now the questions will, in the beginning, reflect the presuppositions that we have already. Very often we ask what we might call a wrong question. I'm not meaning a personal insult to you or anything, but rather that part of the inquiry is to look at the question — to question the question — is it an appropriate question, or does it already contain presuppositions — the same ones which are causing us to question in the first place. So in the inquiry we will go back and forth, to look into our questions and proceed deeper. Now you can see how inappropriate it would be for me just to try to take these questions and answer them.

We are going to have, I hope, a dialogue. A dialogue doesn't mean just between two people; but rather the root meaning of 'dia' in Greek means 'through'. And the general picture it suggests is a stream running between two banks. It's the stream that counts. The two banks merely give form to the stream — the stream is common to the two banks. So there'll be a stream of thought or perception, or some sort of energy flowing between us, unfolding, and that would be the meaning of the dialogue. So I think with that introduction somebody could begin. Can we begin the dialogue by raising your question.

Q: With great trepidation I'd like to suggest that perhaps we don't only make an inquiry on the understanding that something is going wrong. When you said that, I looked at it, and I thought very much about how children ask questions, and I don't know to what extent we can keep that in the adult state; but it seems to me that children ask questions out of the sense that there is a connection, and they want the connections made articulate.

Bohm: Yes. I think that's a good point — that out of the natural curiosity which children have they simply want to understand, and they ask questions spontaneously. Now these spontaneous questions will probably have fewer presuppositions in them than the kind of question that we might ask, but I think that generally people, by the time they are adults, are asking questions from the point of view of what has gone wrong as something challenges them to ask a question. If your mind is very lively, then you ask questions anyway.

Q: I was also thinking of that, because my immediate response was, well, I don't always ask questions if something is wrong. However if I look at it on a wider scale than just the search for knowledge, or the idea that there is a gap there where I don't know something that I want to know, then there's a non-perfection indicated there. So it's not necessarily that something's wrong, but that something's not whole.

Bohm: Something is incomplete. Yes, that would be more accurate; to say that you may sense something is missing, something is incomplete, and you're asking a question, how to complete it, right?

Q: Maybe. Also I was thinking of the way inventions are discovered, like the steam engine was thought out from someone noticing the way steam behaved, and when it's that kind of questioning or pursuing of thought, it's like a means of what is implicate becoming explicate through us. I mean we are conductors perhaps of a creative intelligence.

Bohm: Yes. That is a view that we may feel there is something in us, and we may ask questions to bring it out. I think that gets us into something deeper then, doesn't it?

Q: Or something more shallow?

Bohm: Well, it depends on whether it's really something in us. You see, we may want to just bring out some superficial ideas that we have; but if you have a deeply felt impulse you may want to bring it out — What does it mean?

Q: In terms of dialogue, what cropped up for me immediately when you said that, is I'm going to think, think, think, and you have said what we need to do is go beyond thinking in order that things should flow through us more naturally. So in this context, here and now — this dialogue — what would be the conditions we could create within us for that to happen?

Bohm: I think that we have to create them. You see, if we try to state these conditions we're going to get into difficulty. I think that there is a question of time involved here. If we take time in ordinary experience we have the present, the past and the future. Now the past is gone; it no longer exists except in memory. The future is just expected. Now if you draw a line and put the present as the dividing point between past and future, it divides what doesn't exist from what doesn't exist. And that also doesn't exist. You see, that also raises a question. We see an inconsistency in our thought; that's really what I meant. Then we are led to raise a question; so somewhere there is a presupposition that's wrong.

This linear view of time is an abstraction. It's like a map. You can use this map to guide you, but the map is not the same as the territory. The map may correspond to some features of the territory, but it is never complete, and also it may be distorted. Now actually we can say that thought is of the past — that it takes time to think. By the time you've thought anything it's over. I would like to make a distinction between thinking and thought. Thinking, clearly, takes a lot of time because you've got to work it all out. And if you're dealing with something as fast as the mind, you can't really think it out because it's going faster than you can think. If we were to put conditions on how to establish dialogue, that would have to be by thinking, right? Now these conditions are expressed through thinking and thought.

Certainly we could put some general conditions, like we should be quiet and have the right atmosphere, and so on, but that won't be enough. The point is that after you have been thinking, this turns into thought — that is, what has been thought. Now thought operates very rapidly. If you have been thinking that this fellow is your enemy, then the minute you see him, you see him as an enemy. That is, whatever you have been thinking enters your experience. So therefore, when we experience things, we experience them filtered through thought which is too fast for thinking.

One of the difficulties is that the thoughts contain all sorts of presuppositions which limit us and hold us in rigid grooves. What we have to do is to discover these presuppositions and get rid of them — get free of them. I don't think that we can establish conditions for a dialogue, except to say that we both want to make a dialogue. You see, what would it mean to try to set up conditions? Suppose we said that certain conditions are needed for a dialogue. You could say the condition is to put yourself aside, to be interested in the dialogue — a few things could be said like that, but...

Q: I'm remembering in your book, you were talking about relevance and irrelevance, and saying that what it takes to cease irrelevant thought — thought out of context — is the act of paying attention to that, and maybe what it takes to have a dialogue is the act of paying attention to having a dialogue.

Bohm: Yes, I think that's the way to put it. But if we give attention to this dialogue, giving attention also to what is maybe getting in the way, then we should have the dialogue.

Q: Does that then go beyond thought in that it becomes a motivation?

Bohm: Yes, well, motivation is not necessarily beyond thought. For example, in a detective story there's the motive for the crime, and usually that can be seen as a form of thought — the person thinks, I would like to have something, I need money, I feel insecure, this person has made me envious, or jealous, I have to kill him now he's done this or that. So motivation does not necessarily go beyond thought; but there is a deep intention that goes beyond thought.

Q: I would like to hear you say how you see the nature of an idea. Because an idea may arise as a result of directive thought, or it may come out of nowhere, as it famously does when you're having a bath or something — sitting in a bath. Suddenly an idea which is important for what you have been thinking about, or indeed as a general problem facing mankind at that moment, or facing you at that moment — suddenly, whoomph, the idea is there. I wondered how in the enfolding, unfolding pattern, you would see an idea.

Bohm: Well, an idea I would see as a kind of seed that unfolds. The word 'idea' has a Greek root, meaning basically, to see; the same, ultimately as *eidos*, image. And it seems to have the same sort of root as the Latin, *videre*. An idea is a way of seeing, just as a thought may be a way of seeing, and I think that a new idea appears as a sort of seed deep in the implicate order from which unfolds all sorts of thought. As you apply the idea to your previous thoughts and to experience, then you find the idea unfolding.

Q: How do you see thinking? Are you connecting that in terms of perceiving and aligning perceptions, or... What do you think about thinking?

Bohm: The way I see thinking is that thinking may arise first of all from sense perception. I think Piaget follows it. He shows that there is a sensorimotor thought. That is, a child learns to handle objects and look at them, and a kind of thought develops which is pre-verbal. For example, he may learn to handle an object, and he turns it around, and he turns it around again, and gets the same object back. Now he gets the idea of the group of two, that two operations bring you back to the same thing. Or he may walk to one place, and then around, and come back, and he gets the idea that there is a certain place in the room to which he can always return. And later he gets more complex ideas. For example, if an object disappears behind a screen and reappears later, he may treat it in the beginning as a new object, but later he gets the idea that it's the same object. Now all of this is before he has words. The point is that elementary thought already develops at the level of senses and images; he begins to form images of the world and space and time, and later he abstracts this to verbal thought and builds it up into higher and higher logical structures. Thinking will arise first of all from perception, but clearly it's affected at each stage by past thought. Whatever you have thought is going to affect the next thought, so in thinking, present perception and past thought are fused, as in fact it is in ordinary experience as well.

One of the purposes of thinking is, as I've said, to solve a problem or to raise a question, to answer a question, whatever the source of that question may be. A person comes out with new thoughts, he finds his old thoughts do not cover the situation, and then in thinking he'll begin to work out new thoughts, and that takes time. We may form new images, new combinations of words, new ideas, and so on.

Q: Then do you see ideas as being generated by the individual, or is there such a thing as an idea from the outside being transmitted from, shall we say, some super-mind?

Bohm: Well, that would be difficult to answer; but
certainly ideas pass from people to people, and very
seldom does anybody get, actually, a totally new idea. But
he may get a considerable change, or put old ideas
together, or he may make a considerable change in ideas
which have already been present. So you can already see
that ideas are in constant flow between people in ordinary
ways. Now whether there is some idea coming from
something beyond this, is difficult to say. In most cases
you can see that the foundation of the idea was already
laid in what came before.

For example, if you take Newton's idea on gravitation;
there was a long period of development. If you go back to
ancient times, you have the idea that celestial matter was
very different from earthly matter, and it was perfect —
earthly matter was imperfect — and there were all sorts of
differences. Now, this began to be questioned as people
came out of the middle-ages; and people found all sorts of
ways that this was not so. They said that celestial matter
should move in perfect circles, and when it didn't do so,
they said, circles on top of circles — epicycles. They tried
to save the idea, right? *(laughter)* That, by the way,
illustrates that you have to be ready — an idea must be
vulnerable — you have to be ready to drop it, just as the
person who holds the idea must be vulnerable, I think. He
should not identify with it.

Now Copernicus proposed that the earth was not the
centre of all this, whereas the old idea was that the earth
was the centre going up toward the heavens. And then,
later, Kepler said that the orbits were ellipses; they were
not circles; they were highly imperfect, and you found
much evidence, you see. You found that the other planets
had satellites just like the earth, and that the moon had
very irregular mountains on it like the earth. A great deal
of evidence accumulated suggesting that celestial matter is
not very different from earthly matter. But still there was
a question that nobody asked. You see, I think this is a
major inquiry. Why doesn't an object like the moon fall?
If it's the same as earthly matter — earthly objects fall —

why isn't the moon falling?

You see, the mind establishes two compartments. It has a presupposition. You say the answer's natural, the moon belongs up in the sky because it's a celestial object, and celestial objects don't fall, right? *(laughter)* I mean, every child would feel that way, and therefore he doesn't ask the question. Very few children would ask the question, why doesn't the moon fall. Perhaps they would, but they take it for granted the moon is not falling because it's a different sort of thing. But of course, here you had these scientists with two different views. One was that all matter is similar, and the other was that it is all different. And these two were kept in separate compartments. Now Newton was supposed to have been sitting under the apple tree, and he saw the apple falling, and you could guess that what he asked himself was that if the apple is falling why isn't the moon falling. So that was a new question. You see, he noticed a discrepancy somewhere. That's what I meant — that somewhere something was inconsistent or incomplete or however you want to put it. And then he said, the answer is, the moon *is* falling. There is universal gravitation attracting all bodies to each other. Now that was a sort of flash — an implicit answer. And then he only had to ask why doesn't it reach the ground? Now the reason it doesn't reach the ground is that it is moving in a tangential orbit so that as it falls it also moves away from the earth at the same time, and it stays going around the circle.

So that's probably the way in which Newton came to the idea of universal gravitation. Now you can see the ground was laid in the whole development of a question, and then by simply becoming aware of the question the answer was already there. You see, the new idea was already in the question, right? So you can ask, where does the facility to be aware of the questions come from, and perhaps that's the nature of awareness. I think a lot of this we may discuss after my next talk this afternoon on meaning and the mind, where I am going to go into a view of the implicate order and the mind, and relating mind and matter, where perhaps we could go further into these questions.

Q: So does that tie in with what you call the force of necessity? I was wondering, how do I get things to be explicate from the implicate order, in the sense of the previous question — where do ideas come from? How powerful are we to explicate from the implicate — and all I could find was this term 'force of necessity', and I didn't understand. Could you explain that?

Bohm: This force of necessity is of course a somewhat vague idea in my mind too. If I see an implicate becoming explicate, then I think of explaining it by saying that there is a still a deeper implicate from which arose the force which made it move from the implicate to the explicate — that is, somewhere in your mind there is implicit a question let's say. In the case of Newton there was a question implicit: Why isn't the moon falling? Now nobody asked that question. There was a lack of force to break through the compartments, right? Now from some deeper source came the attention needed to see that that was a question, and I think that force had to break through the barriers of our conditioning — something new.

Normally most of what we do is the result of thought, which is almost like a program of a computer. We need that, because we need all sorts of thought. When you learn to drive a car it's basically through thinking about it that your action really becomes thought, so that you don't have to continue to think about it. It acts rapidly. So you must have thought running a large part of your life, and the hope is that it will be correct thought. If it's not correct thought, then you have to be ready to pay attention and see when it is not correct. The trouble is that this thought tends to run in grooves — perhaps we'll discuss that further — in which it gets stuck. Why it does so is a complex question. It takes something to break through that. Now in an extreme case we would call that power genius — you see, a passion.

For ordinary affairs the force of necessity is largely quasi-mechanical. That's the nature of thought; it isn't exactly mechanical, but it's something like a computer of a very subtle nature. So a great deal of our life is run that way. But somehow there is the power to break through that. Now the word 'necessary' means it cannot be otherwise. The root of the word is *necesse*, meaning it doesn't yield. Now there are two kinds of necessity. First of all, this mechanical thought has a necessity in it which doesn't yield; it gets stuck. But also some greater force dissolves that so it has to yield. We used to have, when I was a child, the question: What happens when the irresistible force meets the immovable object? The point is there is an object that doesn't yield, and eventually it has to yield to a force which however may not be a force of great power in the ordinary sense, but a force which has great subtlety and depth.

Q: It sort of digs out the object from underneath.

Bohm: It sort of dissolves it from underneath, yes.

Q: And makes the idea that it's a solid and spatial object irrelevant.

Bohm: Yes. It brings out the irrelevance of that idea. You see, every idea is limited. Now we don't accept that usually, at least for certain ideas. For example, as we discussed last night, the idea of nationalism is not accepted as limited because it takes precedence over everything else, as you can see in songs like *Deutschland, Deutschland, über Alles*, which means absolutely unlimited, right? Now that idea was the power which drove us to the First and Second World Wars. It was part of the power, anyway; similar ideas prevailed. Hitler brought out many more ideas of this nature — one nation, one people, and so on. Therefore if you take an idea which is absolutely necessary it will generate behind it the force of absolute necessity — the false force of absolute necessity. I think it's wrong for absolute necessity to come from an idea. Every idea must be vulnerable.

Now is there an absolute necessity? Perhaps there is. We have to enquire. But that absolute necessity cannot take the form of any particular idea.

A question which was implicit in all the questions you raised was: What is freedom, right? And: What is the relation of that freedom to absolute necessity? My own feeling is that absolute necessity is the same as freedom. The true absolute necessity is the same as freedom. Now if you don't accept that...

Q: I think it was Spinoza who said that freedom is the understanding of necessity.

Bohm: Yes. It's also creativity. I think that creativity is an absolute necessity — in some sense. But there's also the mechanical necessity which blocks it. Now necessity is a very interesting question. You raised this question in what you sent me. The opposite of necessity is contingency. It comes from the root *contingere*, meaning to be touched — tangent. Absolute necessity cannot be touched. It doesn't yield. What is contingent can yield to pressure from the outside. If we take this chair, it has a certain kind of necessity that it hold together; but it's contingent on certain conditions. If you were to raise the temperature it would start to burn, and it would go to pieces. If you hit it too hard it would break. So it's subject to external contingencies — this necessity. Now normally necessity in science should be limited by contingency, and this contingency in turn is another form of necessity, which in turn has another context in which it's contingent. That is, the ordinary mode of analysis by thought weaves between necessity and contingency. It cannot do otherwise; and that's a necessity.

Now you mustn't accept this or reject this. I'm trying to say, you see, this is part of a dialogue. Don't accept what I say, but consider it, and if you have a question, please raise it.

Q: I was interested in what you were saying when you were talking about *Deutschland,Deutschland, über Alles*, and the idea that one can have a hierarchy, if you like, a hierarchy of

necessity and contingency, or contingency and necessity.
Something that I don't find pleasing aesthetically is the way
that in this world of form, movements which I might find
unpleasant can take on a seeming necessity. Now I was just
hearing, and I don't know if I was hearing it from you, a
possibility of creating a necessity at a lower level which is
smaller than, or contained in, a larger necesssity, but all the
same, manages to sweep around with a force. What would
you say about that?

Bohm: Yes, well we haven't gone far enough yet. You see, I
think in science, or in ordinary thought, we can work only in
the realm of necessity and contingency, and in so far as
scientists would set up laws which they say are absolutely
necessary, then I think that that would tend to lead to the
same problems that you had before with taking the order of
an increasing perfection as absolutely necessary; or taking
the views of the church that certain orders are absolutely
necessary, or the state, or whatever. So as far as thinking and
thought are concerned, it seems that we must have this
interweaving of necessity and contingency. Now this is what
allows room for freedom because whatever necessity there is,
it is contingent. Any of these rigid necessities defined by
thought are contingent — provided you can discover the
contingencies. Now this is one of the ways we make progress,
you see. We may say it's an absolute necessity that man walks
on the ground, but actually it's contingent because he can
also fly given different conditions, and so on. The point then
is that most of our necessities as far as society is concerned —
with the things we don't like — arise from thought. The
product of thinking is thought. *Deutschland, Deutschland,
über Alles* is thought; it's the thought of absolute necessity. So
most of the things we don't like are the result of the thought
of absolute necessity; which however we also do like, or we
wouldn't think of it, right? I think you get a terrific feeling
by singing that song, or the *'Horst Wessel'* song — any of
these songs will give you a tremendous feeling. And that
creates absolute necessity. But it's not true absolute necessity.
This kind of thought is unfolding, you see; it's unfolded, it's

constantly unfolding in our perception and action. If you think *Deutschland, Deutschland, über Alles*, then that's the way you see it, and feel it. It unfolds.

Obviously the thought is enfolded almost like a kind of program. This program, we're not aware of, right? The brain was never set up to be aware of its programs. You see, this is one of the difficulties of human evolution. You see, some people have made a theory. They say that there are really three brains in us: The reptilian brain, the mammalian brain and then the more modern cortex or intellectual brain. And the reptilian and mammalian brains have come to equilibrium and harmony so that, as far as the animal is concerned, it is more or less suited to survival. And when the brain suddenly expanded for some unknown reason in evolution, it was able to think — to produce thought — but it did not have the capacity to see that thought made a program, and that its subsequent actions were determined largely by that program. It did not see the program. It attributed the actions of the progam to the 'self'. You see, if you sing that song long enough, you will say, 'That's me,' right? 'That's what I want.' You can easily see it's a program; but children don't have this idea at all; they sing it every day at school, and after a little while they've got it.

Q: I'm just wondering where emotion comes into all this. Is emotion a necessary product or program of thinking?

Bohm: Emotion and thinking are almost inseparable. They are just different levels of the same thing. It's known that between the intellectual part of the brain and the emotional part there is a tremendous bundle of nerves going up and down. Every thought affects the emotions, and every emotion comes up to affect the thought. No thinking would take place without the emotion of wishing to think. And according to the way that desire is, thought will go in that direction. And emotion is profoundly affected by thought because thought produces images, you see — not the abstract thought but the image and the feeling. Every thought unfolds into images and feelings which operate the same way as perception would. The thought of danger will produce the same emotions as the perception of actual danger. And the same chemical behaviour.

Q: But one can see feelings more easily than one can see thoughts. I mean, anger; it's very easy to see anger coming up.

Bohm: Well, it's often disguised. People often disguise anger as something else; or else they distinguish righteous anger from not....

Q: But you said at the start that one can't analyse thought with thought.

Bohm: That's what I'm trying to say — that we don't see thought because it's a program.

Q: But certainly I have experienced seeing feelings arise.

Bohm: But have you ever experienced the connection between thought and feeling?

Q: I'm working on it. *(laughter)*

Bohm: You see, the point is that you can go back over some previous experience and see that by thinking in a certain way you became angry or afraid, and saying, 'This fellow's behaviour is outrageous; he can't do this to me and push me around,' and so on. You can see that between people; the thought is passing, and the anger is building up.

Q: You can also see it in yourself.

Bohm: You see it in yourself, the same as everybody else.

Q: What's happening then is that we have thought running in a groove. We've just been sort of programmed.

Bohm: Programmed to absolute necessity.

Q: Now we've got to challenge that. We've got to come in and change that into something new. We've got to make some sort of a creative act to break out of this groove. Now where does that creative act come from, and how do we nurture it?

Bohm: Where it comes from would be hard to say; but let me propose — and I'm going to go further into it when I discuss meaning — that there are deeper and deeper levels — more subtle levels of implicate order — and that thought is on a level that is not very subtle on the whole; but thinking may be subtle because thinking may respond to the deeper levels as well as to the thought. So thinking may be a part of perception, if done right; or else it's done mechanically, and then it gives false perception — illusion.

Now I want to suggest, just for inquiry, that somehow in these deeper, subtler levels, thinking is not conditioned by thought. This mind is largely unconscious to us; it's still a part of us, and that's still, I would say, the individual mind at a deeper level. You raised the question of whether there's something beyond the individual mind, but perhaps we could defer that question.

Q: Do you mean that there is some sort of archetypal pool that we can dip our toes in?

Bohm: Well, it's more of a deeper level of enfolded activity — very subtle — which is intelligent, and has energy and passion.

Q: And which is not motivated by the grooves of thought.

Bohm: Yes. The grooves of thought are rather on the surface, you see.

Q: And it filters through them?

Bohm: It doesn't only filter through them, it's got to dissolve them. The reason we're not conscious of this is that these grooves produce such tremendous effects. If you go to a place like Reno, Nevada, or Las Vegas, and we turn on all these electric lights, then you don't see the stars, and you say that all these electric lights are the main thing, and there is no universe. *(laughter)* That's the way people obviously feel at that place, right? *(laughter)* And they blot out the universe. So when you turn off the electric lights, then the universe comes through. At first it seems something very faint, but that faint thing may represent something immense, whereas the very powerful bright thing may represent nothing much.

Q: What we seem to be talking about is a whole process of invisible thought-forms, and somewhere I feel the need of reality. Somehow we can't actually get to reality because we are stuck behind this screen of thoughts.

Bohm: Well, when it comes to ordinary experience, you see how to get to reality. We say the world here is real; it exists independently of thought. Evidently, this chair — I can have enough experience to see, enough intelligence to see that this chair does not depend very significantly on how I think about it, or how you think about it. What I do with it may depend on how I think about it, and that could be changed; but if I don't do anything with the chair, it will go its own way, and my thoughts will go their way. Now that's OK for objects which are fairly separate from us, and which don't change too fast because, you see, our thoughts are behind the reality. This chair is actually changing but in a subtle way all the time. In a hundred or two hundred or five hundred years it would slowly disintegrate. Now that change is unimportant. So for certain categories of process the thought captures enough of the reality so it's a good guide. So we say here that we've got objective reality; we've got a firm grip on it. Our technology has produced tremendous results out of this thinking which takes a firm grip on reality through experiment, through calculation and through reason, and so on.

49

Objective reality

Now when we come to the mind and the sphere of morals and ethics, and so on, and emotions, all that we've done that way, it seems, hasn't got us anywhere. What is the objective reality of myself? Am I an object at all, or what's the difference between subject and object? What is the nature of the reality of myself and my thoughts? The problem arises there. You'll see yourself, or else the civilization, or that the whole of society is in the same situation — that I suppose you are there, and all of us are here, and I may have some presuppositions about all of us. Is there an objective reality which corresponds? If I presuppose the floor will hold me up, the objective reality would support it; but I may presuppose all sorts of things about you or society which have nothing to do with objective reality. Now that is the thing we have to inquire into. How do we proceed into this area?

Even in the sphere of objective reality, if you consider subtle processes like quantum mechanics, similar problems arise. We don't have that sharp separation between things — observer and observed. And also things are in flux, enfolding and unfolding. Also if you consider long intervals of time, your thoughts become inadequate; things change in ways you didn't know about. So the notion of thought grasping an objective reality is limited. Now I think that science tends to proceed from the assumption that that is potentially unlimited; that that's all there is; that it can be grasped that way. But it's clear that if you look at the mind, that notion just hasn't got anywhere.

Q: I'm looking at this example: There is a bed of hot coals in the middle of this room, and I walk across them barefoot. My objective reality would be that I would burn myself badly, but then there are people who can walk across them.

Bohm: But that's not a difficult question, you see. Similarly, if I tried to run ten miles, I couldn't do it. I would probably collapse and end up with a heart attack. But if somebody else does it, they might do it quite easily. You have different people with different capacities, and you may suppose that some people have the capacity to bring their bodies into a different state so they don't burn. I don't want to say yes or no about that. But that creates no serious problem. You see, the question is: Do you have a complete idea about objective reality? The answer is *no*.

Q: The question then is: What is the connection between the subtler levels of the implicate order and objective reality? I think we've taken these grooves, or these surface ripples, to be objective reality and tried to measure them and see how concrete they are. Well, two things: If they weren't there, and the habit of getting into grooves weren't there, then perhaps it would be a completely different world.

Bohm: The objective reality is there. You see this room is here, and so on. Like I said last night, there are two principles — the wholeness of the whole and the parts, and the partiality of the parts and the whole. Now we must adopt both principles but say which one is ultimate. That's where the choice takes place.

Q: Why do you have to say which one is ultimate?

Bohm: Well, if you say neither is ultimate then I think you're going to say that we just have maxims that we apply. It's a sort of indeterminate thing depending entirely on whatever happens. Now you have to say that your world view, the way you look at things, is part of you, just as any other thought is a part of you. It will determine the way you approach the world. One way is to approach the world neutrally and say, I'm not going to adopt either principle finally. What kind of world will that produce? The other way is to say that I'm going to propose that either the major theme is wholeness, or that it is partiality. Now if I don't decide either, if I think passively, I'm deciding that the major theme is partiality, because then I have the two opposing basic principles.

Q: Then there is nothing above just thought-store; and you said that there are things which are likely to be beyond thought. Therefore you don't have to make a choice; you aren't forced into a choice. I still don't see how one is forced into a choice there.

Bohm: Well, I don't say we are absolutely forced into a choice. I'm saying that there may be a good reason for a choice, in the sense that I'm proposing that if you say wholeness is ultimate, you always leave room for the partiality where it is called for. Then your attitude is going to be one of bringing about wholeness ultimately. Whereas if you don't make the choice, then you are saying that there is no reason to choose between partiality and wholeness.

Q: But making a choice is then an act of faith.

Bohm: Yes. I think at some stage everybody makes some kind of an act of faith in the sense faith was defined by Teilhard de Chardin, as just simply that the intelligence holds to certain things — finally to a certain world view. You see, even to take your suggestion would still be an act of faith, that that would be a better approach, right? That is, you cannot avoid making this act of faith. Therefore the question is: Which one should we suggest?

Q: Is it necessary to make a choice between wholeness and partiality? Couldn't we say, wholeness and partiality together as the ultimate?

Bohm: But then you've chosen wholeness, you see — tacitly. Tacitly, that is the choice, that the two together make the whole. You see, I think that's tacitly a choice of wholeness because then we say, wholeness and partiality. As I said, the wholeness of the whole and the parts is your choice.

Q: I'd like to pursue that. Suppose we say 'isness' is ultimate, which includes wholeness and partiality.

Bohm: Yes, then what is 'isness'?

Q: Yes, you can ask further questions, but you don't have to choose between wholeness and partiality.

Q: You could say 'void' instead of 'isness' if you prefer.

Q: I think there is a presupposition in 'isness', actually. I mean there is a presupposition of wholeness in that, at least that's my impression.

Q: I appreciate that, but you see, everything has a presupposition.

Bohm: The question is then: Which presupposition are we choosing?

Q: Yes. I really feel that *you* have to choose between wholeness and partiality, but *I* do not feel that it's necessary. You can transcend the two, and go to the next level, and include the two. You don't have to choose.

Q: Isn't there also a danger in choosing — that once you've made a choice between these two you are stuck with an invulnerable necessity?

Bohm: Well, not really, because in any concrete situation we say that the partiality may prevail. You see, it's merely a sort of general attitude to favour wholeness.

Q: Isn't it a question of trying on a new mind-set? It demeans it to call it a mind-set; but isn't it a question of not saying that the scope of that insight, or the scope of that elevation of wholeness as opposed to partiality, is absolute and irrevocable and will continue forever, but just to bring that up for a moment to see where that takes us? Because we can't know until we've done that where it might take us.

Bohm: Yes. I'm taking as a proposal to inquire into taking the accentuation of wholeness as our approach. All of my thought is in the nature of proposals here, and we could also

53

propose 'isness'; we could try to explore the relative
advantages or disadvantages of these. Now if I propose to
emphasize wholeness, then in any concrete instance you're
always free to discuss partiality. But there is a sort of general
attitude that we are disposed toward wholeness.

Q: Isn't the conclusion — well, obviously there isn't a
conclusion — but if we followed the concept of wholeness
through, would it not come to the Eastern idea that
everything is there, and nothing exists, and in order to keep
our sanity or purpose, or recognize things connected with
order in the world, we then have to take up dualism? This
may be where our friend says, can we accept wholism and
partiality, so that the partiality aspect is the dualism that we
need to learn lessons which lead us towards wholism —
towards oneness. If we could ever get there, then there
would be nothing that we could imagine that existed
separately.

Bohm: Well, we have in that way a dualism between
wholeness and partiality. That is what I wanted to get
beyond. You could say that 'isness' transcends dualism.
One would then want to say that 'isness' is ultimately of
the nature of wholeness, though it contains partiality as a
real contribution.
　Let me try to put it differently. I think that if we try to
look at this as a description of reality it will make no
sense. We cannot discuss metaphysics in this way — that
our metaphysical ideas are corresponding to some object
that they are describing. Rather, we may ask: What is the
use of metaphysical ideas altogether? We don't know the
nature of reality ultimately. What is the point of
describing it or talking about it? Well, one view is to say
there is no point, and a lot of modern philosophers and
scientists say we shouldn't do it. And that's a reasonable
approach up to a point. But everybody has tacit
metaphysical ideas even though he has explicitly
disavowed them. And therefore ultimately that way of
doing it is only going to put you under the control of

whatever metaphysical ideas you happen to have adopted as presuppositons, perhaps in early childhood. So I think it is valuable to explore your metaphysical ideas, to question them, and to propose new metaphysical ideas, and so on.

When we discuss metaphysics — the nature of reality as a whole — we cannot say we are making a picture of this ultimate reality, but rather we have to say that thought is a movement. That is, thinking and thought are a dance of the mind. The question is, what order of dance? Like the dance of the bees which indicates the direction of the honey in the distance. Now in this dance of the mind, I think all of us want to bring about the greatest possible harmony, realizing that a certain amount of disorder and fragmentation is inevitably part of that harmony. The question is, which one will tend to do that? If we say that truth — the pragmatists have said that truth is that which works, but they usually have a rather narrow idea of what they mean by work, in some narrow domain — but if we say that truth is that which brings about an over-all action toward harmony and order, and which we cannot define, as I said last night...

Q: Going on from that harmony, if we ever achieve that perfect harmony, then there will be no movement of thinking.

Bohm: No. I'm not saying we are ever going to achieve it, but it will move in that direction. There is no final metaphysics. I'm trying to say that there are merely proposals. You see, if we say 'isness' is the fundamental, that would also not be a final metaphysics, but it would be open to still more proposals.

Q: I would fear that that too is an assumption, that if we achieve harmony there wouldn't be any more movement.

Q: Well, there are two possible analogies here.

Q: Could you describe what you mean before the

analogies.

Q: That every thing has a purpose — to achieve peace and harmony. That is a belief system, or perhaps what I might think is necessary. Then that leads to stillness and nothingness and the Indian idea of the Gunas, the three aspects of life, perhaps being the movement from one to...

Q: We are getting away from the essential point. I can say that I experience harmony when I run, when I hurdle, and I know I'm experiencing a lot of harmony, but I'm moving.

Q: Like harmony in an orchestra or something? Something's moving, but it doesn't mean that the orchestra's stopped.

Q: The concept of harmony doesn't exist except also in the presence of disharmony.

Bohm: Yes. I would agree with that. What I'm saying is, the movement is — given disharmony — there's a tendency to move to a harmony. That is really what I want to emphasize. And I agree with what you said, that harmony is movement. But the question really is deeper than that because it's a question of whether movement is repetitive or creative. And harmony requires creativity, but that requires some disharmony.

Q: The electron and proton. If we say that electrons are going around a nucleus, they are in disharmony in fact, because they in themselves, if they were being selfish, would be actually trying to get somewhere. I mean, there is a force there. If they ever said, 'Ah, now we've got peace and harmony,' and sat down and rested, there would be no electron. So this is what I understand by disharmony needing harmony.

Q: Harmony doesn't mean there is no tension between the

parts. Tension is almost essential for harmony. It is not stillness; it is more of an active stillness.

Bohm: Of course. Yes. You see, if you take a musical composition, harmony is the harmony of movement, and movement of various themes which have tension between them.

Q: Not if it is a John Cage. *(laughter)*

Bohm: Yes. Well, he's been trying something else; but in general that is the nature of harmony, that its very nature is harmony in movement. However if any pattern of movement is established and starts to become repetitive, then that is a kind of disharmony. At least in certain areas it may be disharmonious; it's part of the harmony of the universe that the pattern of the electron is repetitive; but it's not part of the harmony of human beings to continually repeat a pattern.

Q: I wanted to hark back to a point you made earlier, that we aren't able to pin down the ultimate, and that we have got to have a model of it as a way of looking at it. We make models of the ultimate reality, and I think what we are trying to get at is that we should know we are making models.

Bohm: Yes. We should know we are making models and realize they are not models of ultimate reality, but proposals. They are part of the dance of the mind. They are not models of anything at all really. Rather we are proposing that if we carry out this dance in this way, that the general result will be more harmonious — not in the nature of something static — but rather in the nature of a creative movement.

Q: But that then begs the question of how do we carry out the dance to produce these harmonies.

Kicking the cat

Bohm: Well, you can't answer that question any more than you can answer the question of how anybody carries out any dance precisely.

Q: The nature of the dance is spontaneous.

Q: But at the end of the day you have got to live in the world, and how do these ideas of wholeness help you live more usefuly in the world?

Q: Ah! *(laughter)* I've been tremendously excited and illumined by this so far because for the last five or six years I have been involved in something called The Hunger Project, and the whole point of that is to say that what keeps death by starvation in place is not what we usually identify, but is a mind-set, or a thought pattern — some tacit metaphysical assumption that hunger is inevitable. And the breakthrough, if we can use that expression, is to address that thought pattern — that hunger is inevitable — to break through that, to create the possibility of people thinking — having the thought — that hunger can end. There isn't space in the world at the moment for the thought that hunger can end, although the physical evidence that it could end before the year two thousand is in the...

Q: I was working at a much more down to earth level.

Bohm: Well, it's the same question in everyday life, you see.

Q: You can get all excited about things like this, and then go home and still kick the cat — go home and still kick the cat after becoming inspired.

Bohm: Yes. Well in that case there is still disharmony, and you have to find out what is causing you to kick the cat. You see, there are certain fixed ideas about the cat or something else *(laughter)*. Obviously the cat is representing something else that you would like to kick, right? You see that everyday life is just the place where this question arises. There are all sorts of fixed thoughts which give rise to disharmony, which is not of the creative kind of tension but rather just generally fragmentary and destructive.

Q: We can't rule out pathological sources of disharmony. You needn't look for a metaphysical explanation. People may have disharmony due to emotional problems. We don't need to look for a metaphysical explanation.

Bohm: But the emotional problems may be connected with metaphysics. You see, there is their view of themselves in the world.

Q: It could be pragmatic; it might be infantile.

Bohm: Yes, but that's metaphysical too. The infant has a very strong metaphysics in which he puts himself at the centre of the universe.

Q: If you think that they're separate from the whole, you are going to have problems.

Bohm: Yes. Now say we want to eliminate hunger, and we see that certain metaphysical attitudes are in the way. So we ask a question here: Why do we want to eliminate hunger? In fact many people would say that since I am a separate being, it's of no interest to me — of no great interest to eliminate hunger. So the attitude that people have to each other, as to whether mankind is a whole — whether the ultimate wholeness is primary, or the partiality — is crucial. Now you could say, 'Well, I'll decide on each case according to my convenience whether it's going to be partiality or wholeness.' You could say, 'Well, at this moment, it's more convenient for me to decide on partiality.' So I feel that some sort of choice — not actually choice but perception — that the ultimate wholeness is the primary feature that will lead to greater harmony in the sense of creative harmony, than would the other one. Or I think, even the neutral one. Then you've opened up the question: When will it be? And I think you are saying tacitly — in so far as you are thinking of the good, and so on — you are already saying wholeness is primary because the good, the same as health, is wholeness — sanity is wholeness, holy is whole, and the very word 'good' means whole as well. It has the same root as 'gathering'. There is a tacit notion in all of this that wholeness is in some sense what has ultimately to prevail. Even its 'isness'. Even if you say you are kicking the cat, well, why shouldn't you kick the cat?

Q: Because it produces suffering for the cat and yourself.

Bohm: Yes, but that suffering is a manifestation of fragmentation in your thoughts and feeling. You see, if somebody was really very cruel, he would say, 'I don't care if the cat suffers — this is one of those instances where partiality must prevail,' and he says, 'I'm one thing and the cat is another, so what does it matter what happens to the cat?' You see, somewhere, tacitly, there is a notion of wholeness behind your whole problem.

Q: So basically, kicking the cat is also hurting you because you are not really separate from the cat.

Bohm: Yes, and because ultimately you want wholeness between you and the cat, and not separation.

Q: I can see that, but how to realize that in my everyday life?

Bohm: Well, that I think is a long question. That requires attention to your thoughts. That is, if you feel the impulse to kick the cat, the point is you suspend that impulse. You begin to see that there are thoughts behind it. I don't know what they would be — the thought of frustration, or the thought that something is in your way, or something

— and you have to begin to follow those thoughts.

Q: That is part of my earlier question about anger. I mean, the difference between the feeling and the thought.

Bohm: I'd say fundamentally they flow into each other; they enfold each other. Feeling is implicitly a thought because if you have a feeling this gives rise to a thought, and a thought is implicitly a feeling. We are going to discuss that this afternoon. So the thought that this cat is in the way, or the thought that somebody has done something nasty to me and I need to release — I should do something — will give rise to the feeling. You see, the thought that a shadow is an assailant will give rise to the feeling of fear. So you cannot separate your thoughts and feelings. They are two sides of the same process.

Q: But if you want to actually put these ideas into everyday life, you've got to stop the fault — wanting to kick the cat — actually going into kicking the cat.

Bohm: Well, the first thing is you are not aware of the thought of wanting to kick the cat. You just become aware of the feeling of wanting to kick the cat. We are not generally aware of the thought that is behind it. I'm trying to say that the thought was like a program, and I come back to the notion that when man evolved, he began to think — produce thought — and he didn't realize he was making programs which would have a profound effect on him. The real question is, how to be aware of these programs. That would greatly change everyday life.

Q: We seem to have come to the point where we assume that the universe is ultimately ethical and harmonious. You know that if you kick the cat you have done something wrong by present day standards. But nevertheless, looking at the universe as a whole, is there any particular reason to think that there is that ultimate harmony which we have somehow lost and have to regain?

Bohm: Well, this is a difficult question. One can see harmony and order in the laws of the universe at a certain time, in certain areas. Of course, there is also disruption. Now I don't think we can find this in the universe. You see, the question is: Do we find it in us? Because the universe which we see is seen from the outside; it's only an abstraction which scientists make, and science has made it according to the way we happen to think, and according to the amount of information we happen to have gathered, and so on. The views of the universe have changed all the time, even in the past 20 or 30 years, and might change again. Now one view is to say that we, in some sense, are a manifestation of something deeper in the universe than the things that science is able to show explicitly. And do we find this in ourselves, is the question. This may have more significance than what can be found scientifically.

Q: You see, what our friend there implies is that you only kick the cat if you are pathological. Whereas perhaps kicking the cat is just a natural thing to do, although it may be wrong by our standards. You see, if you hurt the cat, and you hurt yourself, then there is a lot of hurt in you, and perhaps it belongs there, you know. Who says it doesn't?

Bohm: Well, it's a complicated question. You see, when the cat jumps on another animal then that's just getting its meal, but when you kick the cat you cannot see any reason of that nature why you would do it. You are simply inventing something, a conflict in yourself, trying to get rid of it. That is usually what happens. Therefore you are fragmented and you sense that's not a right state, you see. It's also not right to kick the cat because you feel a certain relationship to the cat, and so on. I think it goes deeper than just the thoughts people happen to have on the subject, but the thoughts have a profound effect in making you kick the cat.

But, I think we discover in ourselves a movement or an urge toward harmony, toward the good as we say, which people have almost lost sight of today because everybody is so cynical; everybody is just for himself, number one comes first, and we are just moving on toward catastrophe, and so on. In the past people did accept the notion that there was the good, right? Now I still think it's so, that there is something in us which is the good. It may be hidden by all these thoughts which we get in the newspapers, and so on. There is this flow of bad news. But I think it's important that we have to hold to that good. That's what I meant by faith. I'm not saying that I know that it's true, but it's the approach — the only approach I can see that will make sense and give us any possibility at all. If we don't take this approach then I think we are finished.

Q: I'd like to see your definition of good.

Bohm: Well, that I can't define.

Q: I feel that what is good is what is constructive and facilitating to the whole.

Bohm: Yes. I would agree with that. I mean, if we define it too far it's going to get in the way. We get a feeling for the good, and we can try to express what we mean in a rough way, but if we don't have that feeling for the good then I think we have nowhere to begin. That is, in this case, feeling is deeper than thought, and the thoughts are coming out of the feelings — out of that feeling comes the thought of the ultimate wholeness of the whole and the parts. That is, the thought tends to express that feeling.

Q: From what you've just said, do you mean that we can develop a sort of self-reference to what's good without recognizing the bad?

Bohm: I don't think there is the bad, except we would say that — you see, there is the word 'sin', and apparently there was a Greek word in the bible meaning....

Missing the mark

Q: *Hamartia*, missing the point.

Bohm: Missing the point. Yes, *hamartia* which meant missing the point, missing the mark. Now that got translated as sin. And repentance was *metanoia*, meaning a transformation of the mind, and got translated as pain, right? Penitence, repentance. The point is that repentance is merely to understand that you missed the mark, you see? *(laughter)* Therefore evil is missing the mark, basically. It is confusion, right? Its ultimate source is the kind of confusion I described about thought.

Q: The correct idea in that context is *metanoia* which means not only recognizing that you have missed the mark, but a change in attitude.

Bohm: Yes. I meant to say that. You would understand, not just missing the mark, but you would really understand what it means.

Q: So, what is the idea of guilt?

Bohm: I think guilt is a very destructive idea. There is responsibility, which is correct, but guilt is a very destructive notion. That programs you in a very bad way.

Q: You've just got to know that you have missed the mark?

Bohm: Yes, and not only know it, but understand it deeply, and feel it so that you don't do it again. So if we say there is no guilt then evil arises in human affairs because man did not understand that when thinking he produced thought, which was a program. He is programmed by that, and he misses the mark. You see, that is my suggestion. Man became confused, and gave it a substantial meaning by the word 'evil', and that made it far worse. Because then he said, there is something called evil, which I must overcome. But the brain that misses the mark is trying to overcome the result of what it itself is doing- it's still missing the mark by calling it something else.

Q: By putting it outside.

Bohm: Yes. Putting it outside its real source. The real source is that I'm missing the mark. Suppose I'm an archer, and I miss the mark, and I say some evil spirit made me miss the mark. That would never get me anywhere. So man's misunderstanding of good and evil was one of the principal reasons — that's a metaphysical notion, good and evil — that metaphysical notion was one of the principal reasons for the propagation of evil. It got in the way of people seeing the real source of the trouble. And even created tremendous passion and violence, and added to the evil.

Q: I think it's very important to find out why we keep on doing this. Because I find in my life that when I do something which is like this kicking the cat, it doesn't sort it out. I kick some other cats. It doesn't stop the pattern of doing nasty things; just goes around me like — boom, boom, boom, boom — as well as when I do something that feels really good.

Bohm: Yes. Well, the pattern — the bad pattern — goes that way because it's a program. You see, once the computer has been programmed it repeats the operation in every context, and the question is: Is it possible to change the program? Thinking cannot change it because it's much too fast. Thought cannot change it because it doesn't understand it. Is there something faster still that can touch the program and really change the brain cells which carry the program — iron it out, you see? Now, I think that is crucial; that is the source of what we've been calling evil and sin and so on, and we can see that this was an inevitable problem that mankind would face in evolution once the brain got bigger and able to produce thought, but not knowing what thought does. That is crucial. Now I think that what I say this afternoon will perhaps be relevant to that. Now it's getting a little late unless somebody has a very important question.

Q: I have two. *(laughter)*

Bohm: Make it one. *(laughter)*

Q: Well, maybe you have answered it, but why, if we do sense an ultimate good or whole, and that this is in the process of unfolding, why was there something in us that caused these polluted thought forms?

Bohm: You see, we shouldn't use the word 'polluted'. That is almost the same as the word 'evil', you see. It creates an emotional overtone that is destructive. We must look at evil neutrally and factually. Don't call it evil because the word 'evil' is already loaded; the word 'polluted' is already loaded; so the word is missing the mark. The brain was going to miss the mark because it did not know that it was producing thought which programs you. The programs are not intelligent, and they will inevitably, sooner or later, miss the mark. I mean no machine can be programmed to anticipate all eventualities. Somewhere it is going to do something wrong. And then the attempt to correct it makes it worse because the brain attempts to correct it under the assumption that it is due to some external cause. Its action is wrong, and it gets worse, so evil multiplies if you assume evil. That is, the assumption of evil produces the indefinite multiplication of evil. Therefore a great deal of what has gone on in the world's religions and morality is really missing the mark.

Q: We have a choice all the time about whether we choose to live in a universe in which there is evil. Or we appear to make a choice; I mean our choice is that we want to have evil in it.

Bohm: It's not clear. People don't understand that they are missing the mark, you see. If an archer misses the mark he has no choice; his muscles are conditioned, and his eye, in such a way that he doesn't do it right. Now the choice is not in the matter of whether you miss the mark or not because you can't choose that; but whether you — well, it's really a question of *metanoia* — repentance — you not only acknowledge, but you deeply understand and feel how you are missing the mark; so you stop it. Now the choice is therefore to look in that direction. You cannot choose to stop missing the mark. If you are a bad archer, or a bad whatever, you cannot choose to become a good one.

Q: But you can practice.

Bohm: But then you have to be attentive as you practice to how you are missing the mark. If you don't do that, then you won't learn.

Q: Actually try it again. Perhaps people try, and have to be aware of why... what's going on.

Bohm: Yes. That's it. But the thing that prevents that is that this may interfere with your self-image. You see, you don't like to be aware of it because you have an image of yourself as perfect, or something. You see, that is also another problem that comes in. We all have it, right?

Q: It's interesting to see how in talking about observing by how far we miss the mark, we perceive evil as opposed to our missing the mark. We don't actually look at the target that we are firing the arrow at; we look out into the evil universe for the cause of whatever has come along.

Bohm: Yes. And we also don't notice what we have done to produce this missing the mark. This is the crucial point. We attribute the thing, as you say, to something outside or inside, some mysterious force. We don't notice that some perfectly simple thing we are doing is missing the mark.

Q: Then fighting evil doesn't achieve — doesn't make any good.

Bohm: It adds to the evil, you see. Fighting evil — the very assumption of evil — multiplies evil. That's how metaphysics comes in. Evil is a metaphysical notion and, you see, the smallest child gets the notion of good and evil; he's got a lot of metaphysics.

Q: Would the assumption of fragmentation add to fragmentation? The structure of our discourse is very paradoxical isn't it? We assume wholeness, but we are always talking about fragmentation, and good and evil.

Bohm: Well, we are calling attention to that fragmentation of good and evil which is the source of what we have been calling evil. We have got to call attention to how we are missing the mark. That's why it looks like a paradox. We cannot call attention to how to hit the mark. That happens in a way that you cannot describe. But what you can call attention to is how you are missing the mark, right? Do you see the point?

Q: Well, I feel it's very problematic that you are trying to bring something about and in doing that you have to repeat old habits of talking. The structure of your discourse is fragmentary.

Bohm: How would you say that?

Q: Well, for instance, you introduced a new mode of using language, a philosophical reinterpretation of language, a 'rheomode'. Well, if you would be actually using that, I suppose you would be out performing what you're pointing at, but you can't do that because no one could talk to you on those terms.

Bohm: Agreed. But therefore we have to do the best we can. You see, the present language is highly fragmentary, but it has possibilities for being used correctly — skillfully and correctly. You see, as we use the language we may miss the mark or we may not, right? The present language is not a hopeless confusion, because you have used it to tell me that it is fragmentary. You have some confidence in the language, right? Well, I'm telling you the way in which I have some confidence in the language. Confidence is the same as faith, right? The same root. So the point is that the necessity of this moment forces us to use this language. I hope that eventually we will — human beings will — have a better language. But we've got to start here where we are.

Now this language is a mixture of all sorts of things, and it can be used skillfully and creatively and artistically to say things which cannot be said in the ordinary way, like, take Shakespeare as compared to an ordinary detective story. So the thing is, we have got to work to use this language to communicate properly. And each time that we have not communicated, then we have missed the mark. You could call that a sin if you like; failure of communication is a sin if you like. This is just a word — not that I advocate that — but just to show the meaning of the whole situation. You see, it's tough to communicate in this language, but that is the challenge we face. There is no other way that I can see.

Q: It would be intelligent to analyze why you miss the mark wouldn't it? And wouldn't the answer always be that you had departed from the principle of whatever it actually is? So you are lacking the correct idea of the principle.

Bohm: Well, no. I think it's different from that — that the principle you are given is abstract; it is not part of you. What is part of you is your habits of moving your muscles, which is partly the result of your way of thinking pre-verbally from early childhood. Your nerves and muscles are set in a certain way so that you cannot do other than to miss the mark the way you are conditioned, the way you are programmed. I think that is the way you experience it. You try it, and you miss the mark, right? You try it again, and you miss the mark. Now perhaps with the aid of a skillful teacher, or quite careful attention, you begin to see some of the movements you are making that are causing you to miss the mark. You gradually feel out how you can change those movements, and slowly you come toward the mark. The same is true in communication, or in any other sphere. Therefore you don't condemn yourself for missing the mark and call it a sin, because that will already throw you off, you see? But rather you say, well that is all that happened, and it has some reason which is in me, my conditioning, you see. Everybody is conditioned according to his history. The conditioning is as much muscular as it is ideas and emotions. If you have not been trained in a certain way you will find it quite impossible to do these movements — or if you have never paid attention to it.

Q: That, of course, is the whole movement behind the Alexander and Feldenkreis approach to body work.

Bohm: Yes. For example, in the Alexander method they train you not by words but by actually pushing on you to get you to do the movements. You see, the words could never tell you how to do it.

Q: Isn't there a danger here that every time I kick the cat I say, 'Oh, that is not sin, I have just missed the mark and kicked the cat?' *(laughter)*

Bohm: If you don't mind that you have kicked the cat, there is no problem, you see.

Q: You have diminished the problem because it is no longer evil?

70

Bohm: No. Look. Do you mind it because it is evil, or do you mind it for some deeper reason? You see, generally people call kicking the cat evil, and therefore they say, 'I feel bad because I kicked the cat. If they didn't call it evil, then it wouldn't bother me.' You see, I don't think that's the attitude. There is a deeper feeling that kicking the cat is the result of missing the mark — that I have some emotional disturbance really directed in some other way, and instead I turned it toward the cat, which is missing the mark, right? I mean, that is the thing that is wrong with it. Now if you said, 'I enjoy kicking cats.' *(laughter)* A tiger can say, 'I pounce on the deer and this is my life and there is nothing wrong with it, it's just getting my meal.'

Q: I feel I ought to point out that I do not have a cat. *(laughter)* You see why. *(More laughter)*

Bohm: There's not much time left. I appreciated you writing these questions, and if you want to write them to me again, I would like to see them for tomorrow. But again we will follow the same procedure.

SOMA-SIGNIFICANCE:
A NEW NOTION OF THE RELATIONSHIP
BETWEEN THE PHYSICAL AND THE MENTAL

After a walk through the Cotswold Hills the group gathered again for Professor Bohm's second talk, which introduced some of the most recent extensions of his work.

Professor Bohm: Today I want to introduce a new notion of meaning which I call soma-significance, and also a notion of the relationship between the physical and the mental. This relationship has been widely considered under the name psycho-somatic. 'Psyche' comes from a Greek word meaning mind or soul, and 'soma' means the body. If we generalize soma to mean the physical, the term psycho-somatic suggests two different kind of entities each existent in itself, but both in mutual interaction. In my view such a notion introduces a split, a fragmentation, between the physical and the mental that doesn't properly correspond to the actual state of affairs. Instead I want to suggest the introduction of a new term which I call 'soma-significance'. This emphasizes the unity of the two, and more generally, with meaning in all its implications and aspects. That is, 'significance' goes on to 'meaning' which is a more general word.

In this approach meaning is clearly being given a key role in the whole of existence. However any attempt at this point to define the meaning of meaning would evidently presuppose that we already know at least something of what meaning is, even if perhaps only non-verbally or subliminally. That is, when we talk we know what meaning is, we could not talk if we didn't. So I won't attempt to begin with an explicit definition of meaning, but rather, as it were, unfold the meaning of meaning as we go along, taking for granted that everyone has some intuitive sense of what meaning is.

The notion of soma-significance implies that soma (or the physical) and its significance (which is mental) are not in any sense separately existent, but rather that they are two aspects of one over-all reality. By an aspect we mean a view or a way of looking. That is to say, it is a form in which the whole of reality appears — it displays or unfolds — either in our perception or in our thinking. Clearly each aspect reflects and implies the other, so that the other shows in it. We describe these aspects using different words; nevertheless we imply that they are revealing the unknown whole of reality, as it were, from two different sides.

You can obtain a good illustration in physics for the unbroken wholeness underlying the aspects that are, nevertheless, distinguished, by contrasting the relationship of electrical poles or charges and magnetic poles. Electrical charges are regarded as separately existent and connected by a field; but magnetic poles are not that way. They are really one unbroken magnetic field. That is, if you take a magnet with a north and south pole, you may consider that there is a field going around the magnet from the north to the south pole. You may have seen this illustrated with iron filings.

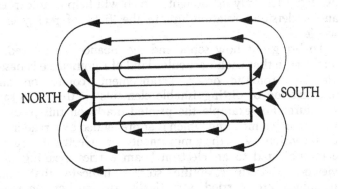

NORTH   SOUTH

Now the point is that if you take this magnet and break it, you get two magnets, each of which has a north and a south pole.

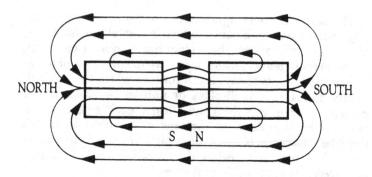

So you can see that there is actually no separate magnetic pole. In fact you may consider that even when it is not broken, every part is a superposition of north and south poles, and you may then understand the relationship as flowing.

With the aid of this concept of opposing pairs of magnetic poles, we can contribute in a significant way to expressing and understanding the basic relationships in the overall magnetic field. I propose to look at soma-significance in a similar way. That is to say, I regard them as two aspects distinguished only in thought, which will help us to express and understand relationships in the 'field' of reality as a whole.

To bring out how soma and significance are related, I might note that each particular kind of significance is based on some somatic order, arrangement, connection and organization of distinguishable elements — that is to say, structure. For example the printed marks on this piece of paper carry a meaning which is apprehended by a reader. In a television set the movement of electrical signals communicated to an electron beam carries meaning to a viewer. Modern scientific studies indicate that such meanings are carried somatically by further physical, chemical and electrical processes into the brain and the rest of the nervous system where they are apprehended by ever higher intellectual and emotional levels of meaning.

As this takes place these meanings, along with their

somatic concomitants, become ever more *subtle*. The word 'subtle' is derived from the Latin *sub-texere*, meaning woven from underneath, finely woven. The meaning is: rarified, delicate, highly refined, elusive, indefinable, intangible. The subtle may be contrasted with the manifest, which means literally, what can be held in the hand. My proposal then is that reality has two further key aspects, the subtle and the manifest, which are closely related to soma and significance. As I pointed out, each somatic form, such as a printed page, has a significance. This is clearly more subtle than the form itself. But in turn such a significance can be held in yet another somatic form — electrical, chemical and other activity in the brain and the rest of the nervous system — that is more subtle than the original form that gave rise to it.

This distinction of subtle and manifest is only relative, since what is manifest on one level may be subtle on another. Thus the relatively subtle somatic form of thought may have a meaning that can be grasped in still higher and more subtle somatic processes. And this may lead on further to a grasp of a vast totality of meanings in a flash of insight.

This sort of action may be described as the apprehension of the meaning of meanings, which may in principle go on to indefinitely deep and subtle levels of significance. For example in physics, reflection on the meanings of a wide range of experimental facts and theoretical problems and paradoxes eventually led Einstein to new insights concerning the meaning of space, time and matter, which are at the foundation of the theory of relativity. Meanings are thus seen to be capable of being organized into ever more subtle and comprehensive over-all structures that imply, contain and enfold each other in ways that are capable of indefinite extension — that is, one meaning enfolds another, and so on. So you can see that the meaning of the implicate order must be closely related. The implicate order is a way of illustrating the way meaning is organized.

In terms of the notion of soma-significance there is no point to the attempt to reduce one level of subtlety in any structure completely to another. For example, if you meet a certain content on one level and then on another, the relationship between these levels is the essential content of yet another level. So it is clear that no ultimate reduction is possible. As the level under consideration is changed, the particular content of what is somatic (or manifest) and what is significant (or subtle) has always therefore to be changing. Nonetheless it is clear that it is necessary for both roles to be present in each concrete instance of experience. You see, it is like the magnetic poles. Wherever you cut the magnet you have a North and a South pole, and wherever you make a cut in experience and abstract something, and say, 'This is the experience,' (which is a bigger context) you have soma-significance. It would be impossible to have all the content on the side of soma or on that of significance.

I have emphasized so far, the significance of soma — that is, that each somatic configuration has a meaning — and that it is such meaning that is grasped at more subtle levels of soma. I call this the soma-signifcant relation which is one side of the over-all process. I would now call attention to the inverse, signa-somatic relation. This is the other side of the same process in which every meaning at a given level is seen *actively* to affect the soma at a more manifest level. Consider for example, a shadow seen in a dark night. Now if it happens, because of the person's past experience, that this means an assailant, the adrenalin will flow, the heart will beat faster, the blood pressure will rise and he will be ready to fight, to run or to freeze. However if it means only a shadow, the response of the soma is very different. So quite generally the total physical response of the human being is profoundly affected by what physical forms mean to him. A change of meaning can totally change your response. This meaning will vary according to all sorts of things, such as your ability or background, conditioning, and so on.

This is different from psychosomatic, because with psychosomatic you say that mind affects matter as if they were two different substances — mind substance affects material substance. Now I am saying there is only one flow, and a change of meaning is a change in that flow. Therefore any change of meaning is a change of soma, and any change of soma is a change of meaning. So we don't have this distinction.

As a given meaning is carried into the somatic side, you can see that it continues to develop the original significance. If something means danger, then not only adrenalin, but a whole range of chemical substances will travel through the blood, and according to modern scientific discoveries, these act like 'messengers' (carriers of meaning) from the brain to various parts of the body. That is, these chemicals instruct various parts of the body to act in certain ways. In addition there are electrical 'signals' — they are not really signals — carried by the nerves, which function in a similar way. And this is a further unfoldment of the original significance into forms that are suitable for 'instructing' the body to carry out the implications of what is meant.

From each level of somatic unfoldment of meaning there is then a further movement leading to activity on a yet more manifestly somatic level, until the action finally emerges as a physical movement of the body that affects the environment. So one can say that there is a two-way movement of energy in which each level of significance acts on the next more manifestly somatic level, and so on, while perception carries the meaning of the action back in the other direction.

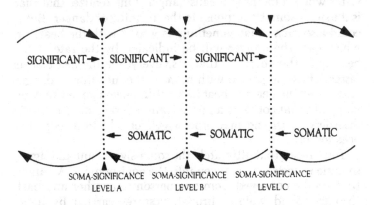

As in cutting a magnet it does not mean that these lines represent distinct levels; they are merely abstracted in our mind.

I want to emphasize here that nothing exists in this process except as a two-way movement, a flow of energy, in which meaning is carried inward and outward between the aspects of soma and significance as well as between levels that are relatively subtle and those that are relatively manifest. It is this over-all structure of meaning (a part of which I've drawn in this diagram) that is grasped in every experience. We can see this by following the process in the two opposing directions. For example, as light strikes the retina of the eye, carrying meaning in the form of an image, the meaning is transformed into a chemical form by the rods and the cones. They in turn are transformed into electro-chemical movements in the nerves, and so on into the brain at higher and higher levels. Then in the other direction, higher meanings are carried electrically and chemically into the structures of reflexes and thus onward toward ever more manifestly somatic levels.

I have been discussing what you might call the normal soma-significant and signa-somatic process. Usually psychosomatic processes are discussed in terms of some disorder, and you can see here that you can also get signa-somatic disorder. For example, normally the heart will beat faster when something means danger. One realizes that that is the signa-somatic response to the meaning of danger. But it could also mean that something is wrong with the heart, in which case the danger will be indicated by the rate of the beating of the heart. In that case everytime the heart beats faster it fills the person with more of the meaning of danger and causes the heart to beat faster still. So you get a runaway loop, and that could be an important component of neurotic disorders — the normal process gets caught in a loop that goes too far.

You can see that ultimately the soma-significant and signa-somatic process extends even into the environment. Meaning thus can be conveyed from one person to another and back through sound waves, through gestures carried by light, through books and newspapers, through telephone, radio, television and so on, linking up the whole society in one vast web of soma-significant and signa-somatic activity. You can say society is this thing; this activity is what makes society. Without it there would be no society. Therefore communication is this activity.

Similarly even simple physical action may be said to 'communicate motion and form' to inanimate objects. Most of the material environment in which we live — houses, cities, factories, farms, highways, and so on — can be described as the somatic result of the meaning that material objects have had for human beings over the ages. Going on from there, even relationships with nature and with the cosmos flow out of what they mean to us. These meanings fundamentally affect our actions toward nature, and thus indirectly, the action of nature back on us is affected. Indeed as far as we know it and are aware of it and can act on it, the whole of nature, including our civilization which has evolved from nature and is still a part of nature, is one movement that is both soma-significant and signa-somatic.

Some of the simpler kinds of soma-significant and signa-somatic activity are just reflexes that are built into the nervous system, or instincts that express the accumulated experience of the species. But these go on to ever finer and more variable responses. Even the behaviour of creatures as simple as bees can be seen to be so organized in a very subtle way by a kind of meaning, in this case through a dance indicating the direction and distance of sources of nectar. Though they might not be conscious of it, there is a meaning going on. With the higher animals this operation of meaning is more evident, and in man it is possible to develop conscious awareness, and meaning is then most central and vital.

In these higher levels this soma-significant and signa-somatic activity shows up most directly. In fact the word 'meaning' indicates not only the significance of something to us, but also our intention toward it. Thus 'I mean to do something' means 'I intend to do it'. This double meaning of the word 'meaning' is not just an accident of our language, but rather it implicitly contains an important insight into the structure of meaning.

To bring this out I would first note that an intention generally arises out of a previous perception of the meaning or significance of a certain total situation. This gives all of the relevant possibilities and implies reasons for choosing which of these is better. Ultimately this choice is determined by the totality of significance at that moment. The source of this activity includes not only perception and abstract or explicit knowledge, but what Polanyi calls *tacit knowledge* — that is, knowledge containing concrete skills and reactions that are not specifiable in language, such as riding a bicycle.

Ultimately it is the whole significance that gives rise to intention, which we sense as a feeling of being ready to act in a certain way. For example, if we see a situation meaning 'the door is open,' we can form the intention to walk through it, but if it means 'the door is closed,' we don't. But even the intention not to act is still an intention. The whole significance helps to determine it. The important point is that the intention is a kind of implicate order; the intention unfolds from the whole meaning. It doesn't just come out of nothing. Therefore a person cannot form intentions except on the basis of what the situation means to him, and if he misses the mark on what it means, he will form the wrong intentions.

Of course, most of the meaning is implicit. Indeed, whatever we say or do, we cannot possibly describe in detail more than a very small part of the total significance that we sense in any given moment. Moreover, when such significance gives rise to an intention, it too will be almost entirely implicit, at least at the beginning. For example, as I said, I have an intention to speak at this moment, and it is implicit what I am going to say; I don't know what I'm going to say exactly, but it comes out. Now the words are not chosen one by one, but rather are unfolded in some way.

Meaning and intention are therefore inseparably related as two sides or aspects of one activity. This is the same as we discussed with soma and significance, and the subtle and the manifest. We are saying that there is one whole of activity abstracted at a certain point conceptually — we make a cut in it — and we say it always has two sides. One of the two sides is meaning, and the other would be intention. But they don't exist separately.

Intentions are commonly thought to be conscious and deliberate. But you really have very little ability to choose your intentions. Deeper intentions generally arise out of the total significance in ways of which one is not aware, and over which one has little or no control. So you usually discover your intentions by observing your actions. These in fact often contain what are felt to be unintended consequences leading one to say, 'I didn't mean to do that.' 'I missed the mark'. In action, what is actually implicit in what one means is thus more fully revealed. That is the importance of giving attention.

To learn the full meaning of our intentions in this way can very often be costly and destructive. What we can do instead is to *display* the intention along with its expected consequences through imagination, and in other ways. The word 'display' means 'unfold', but for the sake of revealing something other than the display itself. As such a display is perceived one can then find out whether or not one still intends going on with the original intention. If not, the intention is modified, and the modification is in turn displayed in a similar way. Thus to a certain extent, by means of trying it out in the imagination, you can avoid having to carry it out in reality and having to suffer the consequences, although that is rather limited.

So intention constantly changes in the act of perception of the fuller meaning. Even perception is included within this over-all activity. What one perceives is not the thing in itself, which is unknown or unknowable, but however deep or shallow one's perceptions, all one perceives is what it means at that moment, and then intention and action develop in accordance with this meaning.

The point is that as you act according to your intention, and as the perception comes in, there can arise an indefinite extension of inward signa-somatic and soma-significant activity. That is, you go to more and more subtle levels and the thing is, as it were, looking at itself at different levels ever more deeply.

Such activity is roughly what is meant by the mental side of experience. When something is going on that is not strongly coupled with the outer physical manifestation of some soma-significant and signa-somatic activity in which it is looking at itself, then we call that the mental side of experience. Now this is only a side. Once again I want to repeat that there is no separation between the mental and the physical. When it gets to the other side where it is primarily concerned with actions it just gets more physical.

Now we can look at this in terms of the implicate or enfolded order, for all these levels of meaning enfold each other and may have a significant bearing on each other. Within this context, meaning is a constantly extending and actualizing structure — it is never complete and fixed. At the limits of what has at any moment been comprehended there are always unclarities, unsatisfactory features, failures of intention to fit what is actually displayed or what is actually done. And the yet deeper intention is to be aware of these discrepancies and to allow the whole structure to change if necessary. This will lead to a movement in which there is the constant unfoldment of still more comprehensive meanings.

But of course each new meaning has some limited domain in which the actions flowing out of it may be expected to fit what actually happens. These limits may in principle be extended indefinitely through further perceptions of new meanings. But no matter how far this process goes there will still be limits of some kind, which will be indicated by the discovery of yet further discrepancies and disharmonies between our intentions, as based on these meanings, and the actual consequences that flow out of these intentions. At any stage the perception of new meanings may dissolve these discrepancies, but there will still continue to be a limit, so that the resulting knowledge is still incomplete.

What this implies is that meaning is capable of an indefinite extension to ever greater levels of subtlety as well as of comprehensiveness — in which there is a movement from the explicate toward the implicate. This can only take place however when new meanings are being perceived freshly from moment to moment. But if significance comes solely from memory and not from fresh perceptions it will be limited to some finite depth of subtlety and inwardness.

Memory, being some kind of recording, necessarily has a certain stable quality which cannot transform its structure in any fundamental way, and has only a limited capacity to adapt to new situations — for example, by forming new combinations of known principles, either through chance or through rules already established in memory. Memory is thus necessarily bounded both in scope and in the subtlety of its content. Any structure arising solely out of memory will be finite, and will be able to deal with some finite limited domain; but of course, to go beyond this, a fresh perception of new meanings is needed. And in fact, when you have a fresh perception you may also see new meanings of your memories. In other words memory may cease to be so limited when there is fresh perception. To go on in this way to new meanings that are not arbitrarily limited requires a potentially infinite degree of inwardness and subtlety in our mental processes. And I am suggesting that these processes have access to an, in principle, unlimited depth in the implicate order.

Thus far I have suggested reasons why meaning is capable of infinite extension to ever greater levels of subtlety and refinement. However, it might seem at first sight that in the other direction — of the manifest and the somatic — there is a clear possibility of a limit in the sense that one might arrive at a 'bottom level' of reality. This could be, for example, some set of elementary particles out of which everything would be constituted such as quarks, or perhaps yet smaller particles. Or in accordance with currently accepted views of modern physics it might be a fundamental field, or set of fields, that was the 'bottom level'. What is of crucial importance is that its meaning would be in principle *unambiguous*. In contrast, all higher order forms in this supposedly basic structure of matter are ambiguous — that is, their meaning is incomplete. There is an inherent ambiguity in any concrete meaning. That is to say, how the meanings arise and what they signify depends to a large extent on what a given situation means to us, and this may vary according to our interests and motivations, our background of knowledge, and so on. But if for example, there were a 'bottom level' of reality, these meanings would be exactly what they were, and anybody who looked correctly could find them. They would be a reality that was just simply there, independent of what it meant to us.

Of course you also have to keep in mind that all scientific knowledge is limited and provisional so that we cannot be certain that what we think is the 'bottom level' is actually so. For example, possibly something other than the present theories will come to reveal a 'bottom level'. But this uncertainty of knowledge cannot of itself prevent us from believing in the existence of some kind of 'bottom level' if we wish to do so. It is not commonly realized however that the quantum theory implies that no such 'bottom level' of unambiguous reality is possible.

Now this is a bit difficult to make clear in this short time, but Niels Bohr, one of the founders of modern physics, has made one of the most consistent interpretations of the quantum theory given thus far, and which has been accepted by most physicists (though few probably have studied it deeply enough to appreciate fully the revolutionary implications of what he has done). To understand this point, first we have to say that while the quantum theory contradicts the previously existent classical theory, it does not explain this theory's basic concepts as an approximation or a simplification of itself, but it has to presuppose the classical concepts at the same time that it has to contradict them. The paradox is resolved in Bohr's point of view by saying that the quantum theory introduces no new basic concepts at all. Rather what it does is to require that concepts such as position and momentum, which are in principle unambiguous in classical physics, must become ambiguous in quantum mechanics. But ambiguity is just a lack of well-defined meaning. So Bohr, at least tacitly, brings in the notion of meaning as crucial to the understanding of the content of the theory.

Now this is a radically new step, and he is doing this not just for its own sake, but he is forced to do something like this by the very form of the mathematics which so successfully predict the quantum properties of matter. This mathematics gives only statistical predictions. It not only fails to predict what will happen in a single measurement, it cannot even provide an unambiguous concept or picture of what sort of process is supposed to take place. So for Bohr the concepts are ambiguous, and the meaning of the concepts depends on the whole context of the experimental arrangement. The meaning of the result depends on the large scale behaviour which was supposed to be explained by the particles themselves. So in some sense you do not have a 'bottom level' but rather you find that, to a certain extent, the meaning of these particles has the same sort of ambiguity that we find in mental phenomena when we are looking at meaning.

This kind of situation is what is pervasively characteristic of mind and meaning. Indeed the whole field of meaning can be described as subject to a distinction between content and context which is similar to that between soma and significance, and between subtle and manifest. Content and context are two aspects that are inevitably present in any attempt to discuss the meaning of a given situation. According to the dictionary, the content is the essential meaning — for example, the content of a book. But any specifiable content is abstracted from a wider context which is so closely connected with the content that the meaning of the former is not properly defined without the latter. However, the wider context may in turn be treated as a content in a yet broader context, and so on. The significance of any particular level of content is therefore critically dependent on its appropriate context, which may include indefinitely higher and more subtle levels of meaning — such as whether a given form seen in the night means a shadow or an assailant depends on what one has heard about prowlers, what one has had to eat and drink, and so on. So you see, this sort of context-dependence is just what is found in physics with regard to matter, as well as in considerations of mind or meaning.

Now I believe Bohr's interpretation of quantum theory is consistent, and he has produced a very deep insight at this point; but it is still not clear why matter should have this context-dependence. He just says that the quantum theory gives rise to it.

However, in terms of the implicate order, an alternative interpretation is possible in which one can ascribe to phenomena a deeper reality unfolding, which gives rise to them. This reality is not mechanical, rather its basic action and structure are understood through enfoldment and unfoldment. What is important here is that the law of the total implicate order determines certain sub-wholes which may be abstracted from it as having *relative* independence. The crucial point is that the activity of these sub-wholes is context-dependent, so that the larger content can organize the smaller context into one greater whole. The sub-wholes will then cease to be properly abstractable as independent and autonomous. The implicate order makes it possible to discuss the notion of reality in a way that does not require us to bring in the measuring apparatus, which Bohr does. He makes the context very much dependent on the apparatus; but he does so by making nature generally context-dependent. That is to say, the situation of any part of nature is context-dependent in a way that is similar to the way that meaning is dependent on its context — that is, as far as the laws of physics are concerned.

That would suggest that in a natural way one might extend some notion similar to meaning to the whole universe. It is implied that each feature of the universe is not only context-dependent fundamentally, but also that the grosser, manifest features depend on the subtler aspects in a way that is very analagous to soma-significant and signa-somatic activity. So something similar to meaning is to be found even in the somatic or physical side.

Now as I said, this holds for *us* both mentally and physically. It would suggest that everything, including ourselves, *is* a generalized kind of meaning. Now I am not thereby attributing consciousness to nature. You see, the meaning of the word 'consciousness' is not terribly clear. In fact, without meaning I think that there would be no consciousness. The most essential feature of consciousness is consciousness of meaning. Consciousness is its content; its content is the meaning. Therefore it might be better to focus on meaning rather than consciousness. So I am not attributing consciousness as we know it to nature, but you might say that everything has a kind of mental side, rather like the magnetic poles. In inanimate matter the mental side is very small, but as we go deeper into things the mental side becomes more and more significant.

All of this implies that one can consistently understand the whole of nature in terms of a generalized kind of soma-significant and signa-somatic activity that is essentially independent of man, and that indeed it is more consistent to do this than to suppose that there is an unambiguous 'bottom level' at which these considerations have no place. I would say that the crucial difference between this and a machine is that nature is infinite in its potential depths of subtlety and inwardness, while a machine is not. Although to a certain extent a machine such as a computer has something similar. So it is in principle possible in this view to encompass both the outward universe of matter and the inward universe of mind.

In this approach, the three basic aspects arise:

Soma
Significance
Energy

Q: Could you just repeat the meaning of soma-significant and signa-somatic again?

Bohm: Yes. Now soma-significance means that the soma is significant to the higher or more subtle level. Signa-somatic means that that significance acts somatically toward a more manifest level.

Now I'm going to look at physical action in a similar way — to say that in the unfoldment of matter there is a kind of soma-significance; that the soma may be significant to a deeper level. So let's say that something unfolds and has a significance, and as a result something else unfolds.

In explaining this I should first discuss the work of the well-known psychologist Piaget, who has carefully observed

87

and studied the growth of intelligent perception in infants and in young children. This led him to say that this perception flows out of what is in effect a deep initial intention to act toward the object. You can see the soma-significance coming in here. This action may initially be based partly on a kind of significance that objects have, which is grounded in the whole accumulated instinctive response to the experience of the species, and partly on a kind of significance that is grounded in his own past experience. Whatever its origin may be, Piaget says, what this action does is to incorporate or assimilate its object into a cycle of inward and outward activity. He moves out, he sees it, he acts on it and that changes his perception, and he acts again. His intention is implicitly in at least some conformity with what he expects the object to be, but it might be vague. The action comes back to the extent to which the object fits or doesn't fit his intention. Then this brings about a modified intention with correspondingly modified outward action. This process is continued until a satisfactory fit is obtained between intentions and their consequences, after which it may remain very stable until further discrepancies appear.

Piaget points out however that the initial intention need not be directed primarily toward incorporating the object into a cycle of activity in order to produce a desired result such as enjoyment or satisfaction. Instead it might be directed mainly at perception of the object. For example, the child may initiate movements aimed at exploring and observing the object, such as turning it around, bringing it closer to look at it, and so on. From such an intention it is possible for him to begin with all sorts of provisional feelings as to what the object might be, and to allow these to unfold into actions which come back as perceptions of fitting or non-fitting. This leads to a corresponding modification of the detailed content of the intention behind these movements until the outgoing actions and incoming perceptions are in accord. This is a very important development of intentional activity which makes possible an unending movement of learning and discovering what has not been known before. So we want to say that this soma-significant and signa-somatic activity, constantly going back and forth, is what is involved in learning. And we can say that this is going on, not only in regard to outward objects, but inwardly — that is, for example, with regard to thought. And there may be another level which picks up the meaning of the thought and takes an action toward that thought while thinking another thought to see if it is consistent. If it is not, then the intention changes until we get a consistent relationship between the thought which arises from the deeper intention and the thought that was first being looked at. You see, you may have a thought that you want to look at, and there may be a deeper intelligence which is able to grasp the meaning of that thought in a broader context and take an action toward it by, as it were, thinking again and seeing whether the thought which comes out is coherent with the thought with which you started. And if it's not, then you can start to change that action until it is. Or you can change the thought. Change can occur at various levels.

So all of these levels of meaning enfold each other and have a certain bearing on each other. This whole process is always soma-significant and signa-somatic going to ever deeper levels. When I talk of these processes I don't only mean going outward into the manifest world, but also the deeper mental processes being explored by still more subtle mental processes. So you could say that the mind has available in principle an unlimited depth of subtlety, and learning can take place at all these levels.

Now what is important is not only what to think but how to think. But if we ask how we think, it may be just as difficult to answer as how do you ride a bicycle? It is at the tacit level of knowing, or at the subtle level, that how to think takes place. You cannot say how to think but you can learn, as I have just been describing, through signa-somatic and soma-significant activity.

To sum up what I've just been saying, a somewhat similar view can be applied within matter in general. So one may think of the whole thing as one process — as an extended idea of meaning and an extended idea of soma. That is, meaning and matter may not have the same sort of consciousness that we have, but there is still a mental pole at every level of matter, and there is some kind of soma-significance. And eventually, if you go to infinite depths of matter, we may reach something very close to what you reach in the depths of mind. So if you consider it, we no longer have this division between mind and matter.

Now we have in this whole process these three aspects: soma and significance and an energy which carries the significance of soma to a subtler level and gives rise to a backward movement in which the significance acts on the soma. Modern physics has already shown that matter and energy are two aspects of one reality. Energy acts within matter, and even further, energy and matter can be converted into each other, as we all know.

From the point of view of the implicate order, energy and matter are imbued with a certain kind of significance which gives form to their over-all activity and to the matter which arises in that activity. The energy of mind and of the material substance of the brain are also imbued with a kind of significance which gives form to their over-all activity. So quite generally, energy enfolds matter and meaning, while matter enfolds energy and meaning.

You can see here how the middle term enfolds the other two:

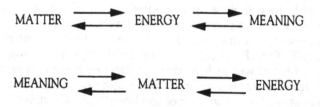

But also meaning enfolds both matter and energy. The way we find out about matter and energy is by seeing what it means.

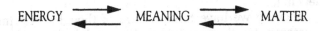

ENERGY ⇄ MEANING ⇄ MATTER

So each of these basic notions enfolds the other two. It is through this mutual enfoldment that the whole notion obtains unity. So we can put all these relationships together:

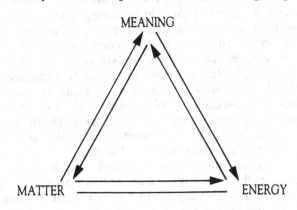

However in some sense the enfoldment by meaning seems to be more fundamental than the enfoldment of the other types, because we can discuss the meanings of meaning. In some sense meanings enfold meanings. But we cannot have the matter of matter, or the energy of energy. There seems to be no intrinsic enfoldment relation in matter-energy. Matter enfolds energy, and energy enfolds matter, according to this view, by way of significance. But meaning refers to itself directly, and this is in fact the basis of the possibility of that intelligence which can comprehend the whole, including itself. On the other hand, matter and energy obtain their self-reference only indirectly, firstly through meaning. That is, we can refer matter back to itself by first seeing what it means to us, and then going back. Or we can refer matter to energy, or energy to matter, by seeing what they mean. We refer them to each other reflexively, but only through their meaning.

Generally we have this problem of thought referring to something else, thus creating division and dualism. Even the thought that the universe is one unbroken whole in flowing movement refers to a universe which is one whole unbroken movement, and beside that there is the thought. So we therefore have two nevertheless. What we would like is a view in which the thought itself is part of the reality.

Usually we think of thought in correspondence with some object; the features of the thought correspond to some object. But as soon as you say a thought corresponds to an object, you immediately have, tacitly, a division between the object and the thought. In reality, we are saying that the thought is a part of the soma-significance and cannot be absolutely distinct from the object. Only in certain limited areas is the distinction useful or correct — that is, where the thought has a negligible effect on the object. This is the area of all practical activity, technology, and so on.

The modern mechanistic approach says that this area covers everything; but what I am saying is that it is a small area within a much vaster field. So we are not denying that kind of thought; we are saying it is only valid in a limited area.

The problem of conceiving of a universe that can refer consistently to itself has long been a difficult one that has not been resolved in a really adequate way. But the field of meaning can refer to itself, and of course, it also presupposes the context of the universe to which it also refers. Meaning, though, has nevertheless been regarded as peculiar to our own minds and not as a proper part or aspect of the objective universe. However if there is a generalized kind of meaning intrinsic to the universe, including our own bodies and minds, then the way may be opened to understanding the whole as self-referential through its 'meaning for itself' — in other words, by whatever reality is. And the universe as we now conceive it may not be the whole thing.

The aspect of soma cannot be divided from the aspect of significance. Whatever meanings there may be 'in our minds', these are, as we have seen, inseparable from the totality of our somatic structures and therefore from what we *are*. So what we are depends crucially on the total set of meanings that operates 'within us'. Any fundamental change in meaning is a change in being for us. Therefore any transformation of consciousness must be a transformation of meaning. Consciousness *is* its content — that is its meaning. In a way, we could say that we *are* the totality of our meanings.

If we trace some of these meanings to their origins, we find that most of them have come from society as a whole. Each person takes up his own particular combination of the general mixture that is available in a society. And so at least in this way, every person is different. Yet the underlying basis is characterized mainly by the fundamental similarity over the whole of mankind, while the differences are relatively secondary. And insofar as man has the capacity to get beyond that, that also is common.

These meanings change as human beings live, work, communicate and interact. These changes are based for the most part on adaptation of existent meanings. But it has also been possible from time to time for new meanings to be perceived and realized — in other words, made real. Perceptions of this kind have generally occurred when someone became aware that certain sets of older meanings no longer made any sense. This may be understood as a vast extension of what happens in the development of intelligence in young children. That is, as they see something about which they are puzzled, they have to see its meaning in a new way.

Now we can say that we are puzzled about the whole of life, and we have to see it with a new meaning. If you look at life as a whole it doesn't seem to make that much sense — the way we live, and so on. The childlike attitude would ask, 'Well, what does it mean?' And some as yet incompletely formed notion of a new meaning that removes the contradictions in the older meanings may begin to penetrate a person's intentions. As I explained, the actions unfolding from the intentions would be displayed, for example, in the imagination, and the discrepancies between what is

93

displayed and what is intended would lead to a change of intention aimed at decreasing this discrepancy, and so on. In this way a greater clarification of the meaning would occur along with a possibility of realizing it through a change in intention, because it is only when one's purpose or intention changes that a new meaning can be realized. Then, often in a flash that seems to take no time at all, a coherent new whole of meaning is formed, within which the older meanings may be comprehended as having a limited validity within their proper context.

Now if meaning is an intrinsic part of not only our reality but reality in general, then I would say that a perception of a new meaning constitutes a creative act. As their implications are unfolded, when people take them up, work with them, and so on, the new meanings that have been created make their corresponding contributions to this reality. And these are not only in the aspect of significance but also in the aspect of soma. That is, the situation changes physically as well as mentally.

Therefore each perception of a new meaning by human beings actually changes the over-all reality in which we live and have our existence — sometimes in a far-reaching way. This implies that this reality is never complete. In the older view, however, meaning and reality were sharply separated. Reality was not supposed to be changed directly by perception of a new meaning. Rather it was thought that to do this was merely to obtain a better 'view' of reality that was independent of what it meant to us, and then to do something about it. But once you actually see the new meaning and take hold of your intention, reality has changed. No further act is needed.

Seeing something intellectually or abstractly, though, will not change your intention. You may say that you need an act of will to change it, but I think that when you really see something deeply with great energy, no further act of will is needed. If you really see a new meaning to be true, then your intention will change — unless there is something blocking it, such as your conditioning, or the 'program'. And if something is blocking it, then the will is not going to help, because you don't know what the block is. Therefore you have to see the meaning of the block. So choice and will are of limited significance — valid in certain areas. But I think something deeper is needed if you are discussing the transformation of mind or consciousness or matter — they really all change together.

You see, the deep change of meaning is a change in the deep material structure of the brain as well, and this unfolds into further changes. Every time you think, the blood distribution all over the brain changes; every emotion changes it. Between thinking and the somatic activity there is also a tremendous connection with the heartbeat and the chemical constitution of the blood, and so on. The new meaning will produce different thought and therefore possibly an entirely different functioning of the brain.

We already know that certain meanings can greatly disturb the brain, but other meanings may organize it in new ways. And when the brain comes to a new state, new ideas become possible. But the new meaning is what organizes the new state. If the brain holds the old meanings, then it cannot change its state. The mental and the physical are one. A change in the mental is a change in the physical, and a change in the physical is a change in the mental. (In fact, there has been some discussion of what is called subtle brain damage in animals in which no physical abnormality can be found; but some disturbance of function takes place when the animals are put under stress. So you see, we could say that living as we do, we probably have a great deal of subtle brain damage. In other words, the brain is damaged at a subtle level that might not show up at the cellular level but deep in the implicate order. Eventually of course, it shows up in the cellular level too.) So instead of saying that when we see a new meaning we make a choice and then act, we say that the perception and realization of the new meaning in our intention is already the change.

This point is crucially significant for understanding psychological and social change. For if meaning is something separate from human reality, then any change must be produced by an act of will or choice, guided perhaps by our new perception of meaning. But if meaning itself is a key part of reality, then, once society, the individual and their relationships are seen to mean something different from what they did before, a fundamental change has *already* taken place. So social change requires a different, socially accepted meaning, such as in the change from feudalism to the forms that followed it, or from autocracy to democracy, or to communism, and so on. According to the meanings accepted, the entire society went.

These meanings may have been correct or incorrect. But once the meanings become fixed, the whole thing must gradually go wrong. Or to put it differently, what man does is an inevitable signa-somatic consequence of what the whole of his experience, inward and outward, means to him. For example, once the world came to mean a set of disjointed mechanical fragments one of which was himself, people could not do other than begin to act accordingly and engage in the kind of ceaseless conflict that this meaning implies. The meaning of fragmentation includes conflict and self-centredness — in other words, not creative tension but meaningless conflict.

However if mankind could sustain a perception and realize this perception signifying that the world is an unbroken whole with a multiplicity of meanings, some of which are fitting and harmonious and some of which are not, a very different state of affairs could unfold. For then there could be an unending creative perception of new meanings that encompass the older ones in broader and more harmonious wholes which would unfold in a corresponding transformation of the over-all reality that was thus encompassed.

Here it is worth noting that our civilization has been suffering from what may be called a failure of meaning. Indeed from earliest times people have felt this as a kind of 'meaninglessness' of life. Whether this is more prevalent today, I don't know, but people say it is. But in this sense, meaning also signifies value. That is to say, a meaningless life has no value; it is not worth living. But of course it is impossible for anything to be totally free of meaning. For as we have explained earlier, the notion of generalized soma-significance regarded as valid for the whole of life, implies that each thing *is* its total meaning — which of course must include all of its relevant context. What I intend by 'meaningless' therefore is that there *is* a meaning, but that it is inadequate, because it is mechanical and constraining and hence of little value and not creative. A change in this is possible only if new meaning is perceived that is not mechanical. Such a new meaning, sensed to have a high value, will arouse the energy needed to bring a whole new way of life into being. You see, only meaning can arouse energy.

At present people don't seem to have the energy to face this sea of troubles that threatens to overwhelm us, generally speaking. If we take a mechanical meaning, it tends to deaden the energy so that people remain indefinitely as they have been, or at best allows change in limited directions such as the continuation of the development of technology, and so on. So I am saying that meaning is fundamental to what life actually *is*.

Now you can extend this to the cosmos as a whole. We can say that human meanings make a contribution to the cosmos, but we can also say that the cosmos may be ordered according to a kind of 'objective' meaning. New meanings may emerge in this over-all order. That is, we may say that meaning penetrates the cosmos, or even what is beyond the cosmos. For example, there are current theories in physics and cosmology that imply that the universe emerged from the 'big bang'. In the earliest phase there were no electrons, protons, neutrons, or other basic structures. None of the laws that we know would have had any meaning. Even space and time in their present, well-defined forms would have had no meaning. All of this emerged from a very different state of affairs. The proposal is that, as happens with human beings, this emergence included a creative unfoldment of generalized meaning. Later, with the evolution of new forms of life, fundamentally new steps may have evolved in the creative unfoldment of further meanings. That is, we may say that some evolutionary processes occur which could be

97

traced physically, but we cannot really understand them without looking at some deeper meaning which was responsible for the changes. The present view of the changes is that they were random, with selection of those traits that were suited for survival, but that does not explain the complex, subtle structures that actually occurred.

The question is how our own meanings are related to those of the universe as a whole. We could say that our action toward the whole universe is a result of what it means to us. Now since we are saying that everything acts according to a similar principle, we can say that the rest of the universe acts signa-somatically to us according to what we mean to it.

These meanings do not all fit harmoniously, but if we are perceptive of the disharmony, we may continually be bringing about an increase in harmony. That is to say, there is no final meaning or no final harmony, but a continual movement of creativity — or of destruction. In the long run, only those meanings which allow changes that tend to bring about accord between us and the rest of the universe will be possible. We can say that that is true for the universe as a whole, and that nature is experimenting with all sorts of meanings. Some of them will not be consistent, and they will not survive. So anything that has survived for quite a long time is bound to have a tremendous degree of coherence with the rest of the universe.

We are proposing that this holds for both living beings and for matter in general. We may say then that the harmony is never complete and cannot be so. Even now a further creation of meaning is going on in a process that includes mankind as part of itself. Not merely man's physical development but a constant creation of new meanings that is essential for the unfoldment of society and human nature itself. Even time and space are part of the total meaning and are subject to a continual evolution. As I indicated, at the beginning of the 'big bang', time and space did not mean what they now mean. In this evolution, extended meaning as 'intention' is the ultimate source of cause and effect, and more generally, of necessity — that which cannot be otherwise.

Rather than to ask what is the meaning of this universe, we would have to say that the universe *is* its meaning. As this changes, the universe changes along with all that is in it. What I mean by 'the universe' is 'the whole of reality' and what is beyond. And of course, we are referring not just to the meaning of the universe for us, but its meaning 'for itself', or the meaning of the whole for itself.

Similarly there is no point in asking the meaning of life, as life too *is* its meaning, which is self-referential and capable of changing, basically, when this meaning changes through a creative perception of a new and more encompassing meaning.

You could also ask another question: What is the meaning of creativity itself? But as with all other fundamental questions we cannot give a final answer, but we have to constantly see afresh. For the present we can say that creativity is not only the fresh perception of new meanings, and the ultimate unfoldment of this perception within the manifest and the somatic, but I would say that it is ultimately the action of the *infinite* in the sphere of the finite — that is, this meaning goes to infinite depths.

What is finite is, of course, limited. These limits may be extended in any number of ways but, however far you go, they are still limited. What is limited in this way is not true creativity. At most it leads to a kind of mechanical rearrangement of the kinds of elements and constituents that are possible within those limits. One may think of anything finite as being suspended in a kind of deeper infinite context or background. Therefore the finite must ultimately be dependent on the infinite. And if it is open to the infinite then creativity can take place within it. So the infinite does not exclude the finite, but enfolds within it and includes and overlaps it. Every finite form is somewhat ambiguous because it depends on its context. This context goes on beyond all limits, and that is why creativity is possible. Things are never exactly what they mean; there is always some ambiguity.

I think maybe that's where we'll end this and open it up for discussion.

Q: Are you saying that the ultimate, deep meaning of something is not subject to the soma-significant or signa-somatic process? That it just is?

Bohm: Well, that's one way you can think of it — that the signa-somatic and soma-significant process is the unfoldment of the deep meaning.

Q: I think that the implicate order is intrinsically self-referential. I think it's the result of chaos theory, for example, where you have functions where $xi + 1$ would be a function of xi. We can't equate them, and you get either stable patterns which recur, which will be kind of stable patterns of stable particles; or at certain parameter values you get chaotic behaviour. This very much would seem to me to be one model which you could use to explain relatively stable patterns in this over-all movement or field, and relatively chaotic things — unexplainable things to our minds. Because chance events, or so-called random events, and strictly deterministic events seem to be coming from the same source.

Bohm: I would say that there is no disorder, but that chaos is an order of infinitely complex nature. One form of chaos is entropy but, you see, there may be chaotic forms that are not in equilibrium.

Q: Some method in madness.

Bohm: ...and there's sometimes madness in method, right? *(laughter)*

Q: If the form of things — this planet with its mental substance and its physical substance — is entirely dependent on what's happening, what meanings it...

Bohm: I would say 'ultimately', not 'entirely'.

Q: That puts man in an interesting position. It would seem that human consciousness has the ability to contain meaning, and to transform meaning in a way that...

Bohm: To *be* meaning — in a much more full way than, say, ordinary, inanimate matter.

Q: The question is: Is the physical and mental realm which we consider to have an existence of itself, is it in fact dependent on man for its existence, because man provides the meaning?

Bohm: Well, I propose that it isn't. There may be something of a signa-somatic and soma-significant nature going on that has meaning for itself, or for the deeper levels — the ever more subtle levels. Nevertheless man may reveal something. You see, there's an argument given by Teilhard de Chardin saying that if you take the present views of science, they explain everything except man — his consciousness and what goes wrong with it, and so on. Now one view is to say that's a minor aberration, and eventually perhaps we'll overcome that. But the other view is to say that the thing unexplained is the sign of something much deeper. It often happens in science that something that cannot be explained, however small it is, shows up or reveals something much more fundamental. Like a few uranium atoms breaking up revealed a tremendous possibility. You could say, well it's just a few uranium atoms, what does it matter? So therefore man may be revealing something much deeper in the meaning of reality, because his brain is capable of a much greater unfoldment of meaning.

Q: It seems we have to assume the possibility of that to actually reveal that possibility.

Bohm: Yes. Well that is a thing we sometimes have to do, to presuppose something in order to realize it. That is, you can see its possibility and presuppose its reality, in order to bring about its reality. Now that is a proposal which we are going to explore. We are saying that everything that we're doing

101

here is a proposal that we are exploring. We are not starting out and saying this is the way it is.

Q: The proposal that you are offering, if I'm understanding you correctly, is that your thinking is part of my consciousness, and then becomes part of my thinking which changes my thinking, or modifies it.

Bohm: And vice versa.

Q: ...and vice versa. So we establish a dialogue here in consciousness. And you are suggesting that matter is behaving similarly, in different levels, according to the range of its conscious awareness, whatever that might be.

Bohm: Yes. But I would say according to the range of meaning. I think conscious awareness, its essential feature, is meaning. The content that one is consciously aware of, is meaning. And that meaning is active. The activity of consciousness is determined by the meaning. Therefore you could say that consciousness, both in the features that we experience and in its activity, is meaning. Without meaning there is no consciousness. And the greater the development of meaning, the greater the consciousness.

Q: And vice versa.

Bohm: Yes.

Q: So we have to find what meaning is.

Bohm: Well, I don't think we can because that still would presuppose meaning. You see, meaning is capable of indefinite extension. And regarding this as a proposal that is to be explored — that is, if meaning is what life is — then a change of view of meaning is a change of life — a change of our lives — if we could really realize it.

Q: When you say that our consciousness is meaning, then we

have got to have some understanding as to what we mean
by meaning, or else we can't actually go on talking.

Bohm: Well, I've sort of unfolded some of the things I
mean by meaning — that there is a perception of meaning
going from one level to another, and a signa-somatic
activity backward. But meaning is this whole activity in
which the meaning of the soma comes into the next level
of subtlety, and action goes back out. Those are some of
the essential aspects of meaning — that meaning pervades
being. I think that you cannot get the whole meaning of
meaning because our ideas of meaning are never fully
defined, and we can only sort of develop this meaning.

Q: Can you distinguish between meaning and purpose
here, because meaning is having to carry the burden of the
meaning of purpose.

Bohm: Well, I think purpose is part of the meaning —
that is, our purpose, our intention — I mean to do it. I
want to say that the purpose flows out of the meaning,
and through the action carried out, the meaning further
changes, and we are back in this cycle. So meaning and
purpose flow together.

Q: Still, they should be different, otherwise there wouldn't
be these two words.

Bohm: Well, I was about to say that the difference is like
the difference between the north and south poles of a
magnet. They are introduced for the sake of thought but
there is not actually a sharp distinction between them.
They are two sides of one flow.

Q: As an absolute minimum, can I suggest as a proposal for
us to work on, to define meaning as an appreciation of a
connection between the points involved, which we are now
talking about as meaning. The connection between these
which makes sense — a connection in terms of which we can

correlate those things we are feeling. It might be a minimum requirement.

Bohm: Yes. We have, let's say, a connection between soma and significance, and the two directions of flow.

Q: But then that is passing the things onto the word 'sense'.

Bohm: Sense is meaning, right? It's another level of meaning. You see, I'm trying to say that you cannot get out of the field of meaning. We're unfolding the meaning of meaning that way, but it is still in the field of meaning because its emphasis is meaning. As you say, it's a coherent meaning. But I think the most fundamental things cannot be defined; we can unfold them, but we can't define them.

Q: Would the word 'relationship' come in here? Because it only makes a sensible meaning if all the aspects are related.

Bohm: Yes, if they are related consistently, right? That is, we find that things have no meaning when they are not consistent with each other.

Q: I'm interested in knowing where the word 'energy' belongs in this process, if 'process' is the right word. You're suggesting that in order to move to deeper levels of meaning, or to become aware of all meaning, requires an increase of energy. But presumably 'energy', if we are looking at meaning as being fundamental, is a product of the meaning.

Bohm: Yes. They work together. You see, again, these are three aspects of one whole flow. We distinguish them for the sake of thought in order to show relationships, but we do not start by assuming that they exist distinctly and separately. We have to say something which I didn't say before, that the question we could raise here is between appearance and reality.

Now we usually say that we have the appearance of things which shows up in the senses, and then we have our deeper thoughts which give the esssence. That's one view. But if you look at it carefully, you will see that these thoughts are also appearances. They are limited, and so on. So you could say that the relationship between these two appearances gives a deeper reflection of reality than the one alone. The same, you see, as with all these meanings, right?

Q: What I'm looking at is how one, in fact, conjures the energy - the direct experience to move from one level of meaning to the other.

Bohm: Well, I think it either happens or it doesn't — that if somebody sees something of great meaning, the energy will arise.

Q: In your triangle, where would you see man? I mean, do you see our function being to reveal meaning? Where do we stand, is it that we assume the place of revealing meaning?

Bohm: Well, we reveal meaning in a way which the rest of matter does not. But we are meanings. You see, I think that if you say 'reveal meaning', it suggests that meaning is something different from being.

Q: When I think about the alienation that people feel — sympathizing with all that you have been saying — the message to me is that we can create our own reality.

Bohm: Well, there are some limits to it, but you see, we can do so to a large extent. We don't really know what reality is.

Q: We can define its meaning.

Bohm: Yes, well, then the question is whether it's a coherent reality or not. You see, in previous periods people created a reality according to mythological meanings which had a certain coherence; the action flowing from it produced a certain coherent society, up to a point, and therefore people lived in that reality, you see. We have now created a different reality, and we could still create a yet different reality.

Q: And the other pieces are there when we say we determine a value — a meaning that is important to us. So we set about creating that as part of our life, and then it gets blocked, and then we look within to what it is about us and our meaning that is blocking us, and we change that.

Bohm: Yes. You see, we are stuck with the old meanings which are in a rut — the old thought, the old program. Now insofar as we look into them, then that may change; the block being removed, new meanings can be perceived.

Q: ...and then exchanged. But we are doing it, and that's an important...

Bohm: Yes. We are doing it. We are the doing of it as well.

Q: Then in the picture — the triangle itself — is that a description of the physical universe? And as Graham asks, 'Where does man fit in there?' It made me think, well, is that a description of man and of the physical universe? And what I started to mention earlier is that the flow in the three Gunas — the flow between the three aspects of life — will actually create the physical universe?

Bohm: Well, one could say that this was actually a way of looking at the flow that creates the universe and creates us. It is a way of thinking about it.

Q: Would it be right to say that you would define evolution as the movement of the universe towards its own coherence, in terms of what it is or can be?

Bohm: Yes. Evolution is the unfoldment of these meanings.

Q: In terms of greater and greater coherence?

Bohm: Yes.

Q: And subtlety?

Bohm: Subtlety, yes. *(long pause)* Now what this suggests then, is that we have open to us a tremendous possibility. And in fact, it's the present meanings that are programmed, that are leading us to all these troubles. Now you see, the fact is that the world means a place where everybody must scramble for himself, and where each nation must defend itself, and so on. And that makes inevitable all the consequences that are building up. Now that's the result of history. A change of meaning is necessary to change this world politically and economically and socially. But that change must begin with the individual; it must change for him, right?

Q: The question is finally: Do we have any choice in the way that change goes?

Bohm: Well, can Newton choose to perceive the law of gravitation? You see, I think we can have this passionate intensity to look at it, to explore it, to find out where it isn't making sense, and why, right? Then it will change creatively. You see, we can't *produce* the change that is really needed to change the future of mankind.

Q: Inherent in that whole process is the fact that meaning will unfold as meaning unfolds. I mean, you can't know ahead.

Bohm: No. And that is true in every creative act. I think therefore the attempt to make a plan to change society is not going to work, because society is the result of what it means to us, and the plan will not change what it means. You may plan the perfect socialist society in which there is complete justice and equality, but what the world means to these people is more or less the same as before you made the plan, and it will produce the same sort of society as we see now.

Q: So what's stopping this? Blockage, you said. If it is, how do we unblock the blockage?

Bohm: Yes, well, that's a difficult one....*(laughter)*. You see, to begin with, it requires some attention to reveal this blockage, and to feel it, and to see the thought that is behind it — the fixed thought. Now you see that thought acts like a program which is very fast, and it hides itself. That is the whole problem — that thought tends to conceal itself — as you may have noticed. It's a program which is programmed to conceal itself. Any attack on the program will seem to be an attack on what I am, right? And therefore, automatically, I will prevent any move to make that attack. And concealment is one way of preventing attack.

Q: But there can be other clues — the body, and then the emotions.

Bohm: Yes. The things show up in the body and the emotions, but the connection between that and thought tends to be concealed.

Q: There is an indication that there is a way of removing these blocks, and that that can be a very natural experience. Are we saying that? That by just seeing — the act of seeing our programs — if we can bring ourselves to that state, and consistently, there is an evolution of the universe and the meaning of which we are a part.

Bohm: Yes. If we can see these programs, then they will change — if we see the meaning of these programs as programs. At present they don't mean programs to us; they mean something much more fundamental. They mean *me*, right?

Q: The act of seeing reveals them as programs and defuses them, or disempowers them.

Bohm: Yes. Now the point is to realize that perception.

Q: You might say the blocks are ignorance, and the solution is the light of education.

Bohm: But the difficulty there, is that the education is being provided by people who are still blocked. You see, we need a creative act.

Q: It's more than just seeing it; it takes replacement.

Bohm: It takes realizing it. It's really sticking to that perception and carrying it out. Now that may be very difficult because the thought is very elusive, and it continually creeps in by the back door. But I'm just pointing out the nature of the difficulty that we are facing. Now you could say that you must start with right relationships in your ordinary life and see the blocks to that. But if you start out establishing those relations, you will find difficulties. They are not merely to be overcome, but one has got to see the source of these in the program of thought which includes the emotions and the body. And you see, one can stick to it. There's room for both a flash of insight, and also for the hard work required for continual development. To realize that insight may require quite hard work.

Q: In other words, actually the clue about the thought processes is quite obvious if you look at the interactions in our world. I mean, it's not that difficult, even though thought is programmed to conceal itself. Unless we are grossly dishonest, we can't avoid the fact that we are having a row with someone, or that there is nuclear war.

Bohm: Yes. Now one interesting thing to do is, as the impulse to anger, or to do something develops, if you suspend it then it becomes clear that you have got an impulse to do something that doesn't make sense. You see, if you don't suspend it, but carry it out, then you are carried away, right? Now instead, you could act out the impulse, rather than carry it out — act it out within. That's really what imagination is, right? You see, imagination is not just making a photographic image or a television image, but it's acting out the meaning.

Q: So if you hold back on intention, that will reveal...

Bohm: It helps to. It helps reveal the meaning behind it — the thought behind it with its meaning. Now the intention to strike out verbally or physically when somebody does something outrageous, or insults you, and so on, if you hold it back, then you'll begin to get a feeling in your body of that action flowing out from some deep intention. But behind that intention is the thought, 'He can't do this to me; he's pushing me around; he's acting outrageously; he's violating my rights,' and so on. That is, somewhere you'll find a thought.

Q: If you follow through with the intention, you'll never notice.

Bohm: Yes. If you carry it out, the thought must ordinarily act itself out in order to reveal what it means inside. At a certain stage it's like an actor carried away by his role, and he starts to carry it out rather than act it out. That is, an actor plunging a dagger into you doesn't actually go all the way, right? If he got too engrossed, it might go all the way.

Q: There seems to be a great push for us to want to be spontaneous, which is actually going along with intention.

Bohm: With the program. You see, spontaneity — true spontaneity — is difficult to distinguish from the program, right?

Q: In this area, between the program and this act of spontaneity, is there anything that we can do other than educating in terms of the programs, to train ourselves as an organism, to be able to see further and further, deeper and deeper, to find platforms for ourselves from which we can see?

Bohm: Well, I think that we can first of all try to observe our own habits and programs, and suspend them sufficiently to get some insight; and secondly, in relationship, we can observe this happening. Also, I think, in dialogue it could perhaps begin to be revealed because different people can see different aspects of this whole thing. Now that would require however, as Peter was saying, that people be really friends with each other, so that they can take criticisms from each other — things that will appear to be criticisms. You see, the minute that somebody is pointing out that you are doing something wrong, or silly, or something, then there is a reaction, and you say no it's not. You defend yourself. That is part of the concealment. It's very hard to be able to take this. But in this dialogue it must be possible for it to go both ways, you see? Now say, in the family, the parents will do this with their children, but the children are not allowed to do it with the parents. Usually they say that the children really do not understand enough. But that often conceals the fact that the parents don't understand — don't want themselves criticized or anything. So the question is whether we can establish a relationship truly between two friends, so that this dialogue is really possible?

Q: In addition, there is one's dialogue with oneself. Would meditative and reflective techniques short out the program?

Bohm: Well, we would have to say what they were to see how they get at the programs. The word 'meditation' has so many meanings, and I would say that the attempt to watch one's own programs is the beginning of a kind of meditation, right? But then we could say the attempt to do this in a dialogue would be a kind of social meditation. Now here the notion of the implicate order would be relevant, because we could say there is a flow back and forth — what I am is enfolded in you, and what you are is enfolded in me. And that comes back to me through you, right? Therefore instead of reflecting it within myself, I reflect it in the dialogue; we reflect it in the dialogue. Now that may bring out many things more clearly. So we could say that there is also a social meditation possible; the dialogue could be regarded as that.

Q: Sometimes we are inclined to look for a program that would help us to be aware of our programs. It might be one hour of a certain meditation every morning, or from now on I contact my friend every evening — and this would not be sufficient.

Bohm: No. I think that we can't lay down a program. We have to be creative. The same trick doesn't always work twice. It may work a few times but not necessarily indefinitely.

Q: At best it would set up another level of mechanism.

Bohm: Yes.

Q: And the energy would be reduced as you suggested, because it becomes increasingly mechanical — I meditate every day at 10 am, and so on.

Bohm: Now you see, meditation can take place in any context.

Q: What I find very important about this, right now, is that, although I have come across the thought of creativity coming from nothing, I have never been so aware of it as right now. Because I think I always thought creativity came from spontaneity. And I have just got the point that you were making about spontaneity, and that makes me realize how by standing back and not being spontaneous necessarily, in a difficult situation, you get into a state of knowing what the thought was, and then by ridding yourself of the thought, you are into nothing, like you don't know about anything, and you then can be creative.

Bohm: Yes. That is, true spontaneity may be the suspension of what appears to be spontaneity, but is not truly. You see, the concealment of the program is that it conceals this

112

activity as the appearance of spontaneity, which is not true, right? There are all sorts of clever ways in which the programs conceal themselves as the non-program.

Q: Isn't a very important requirement for change, beyond self-examination, beyond dialogue, that there is a climate, a climate that one finds oneself in?

Bohm: Well, that's true, yes.

Q: Isn't it possible to at least suggest that perhaps the evidence of us here, is that there's a change of climate going on; that perhaps there is a situation where it's maybe created by us, where the climate on earth can become more and more suitable for the sort of change we see as being necessary.

Bohm: Yes. We need a certain environment, a certain place in which we can do this. And people who have the intention, the serious intention to do it, are coming together and creating the opportunity to do so.

Q: There is something here, touching back to what we were saying about the whole being more than the sum of its parts, that in dialogue one can see things if one is willing to be seen, and to take somebody else's perceptions. One can see more with another than one can see oneself. And that we together, collectively, can see more individually. And that works more broadly, in that perhaps, as we are a part of a culture, we can have some sense of where we are going, and get perceptions that aren't available to the individual, or the individual organization, or the individual set, or group, or whatever.

Bohm: Yes, well, we have to explore. You see, I am not telling you that I know how to do it. I've merely said that these are some of the things I've seen, and we've got to take it from there, right? Now the difficulty with our culture is its terrible confusion, right? I mean, I shouldn't use these loaded words, but I think that we hardly even have a culture now. If you take ancient Greece, and, say, a place like Athens, with a small number of people who were not wealthy, and take what they produced; and then take what has been produced by us who are enormously greater in number and far wealthier; even then, most of our good culture, our really good culture, has been done in the past, and that's not a lot. Conditions are not favourable now, because of the general deterioration going on. I think those who feel differently have got to somehow get together, as we are doing now, and create an environment, an ambience, in which we seriously have the intention of exploring being creative in this new way.

Q: You know, there is a funny thing happening, that because of this confusion it seems that there might be some mechanism that culture can use to repair itself and jump to a higher level. There is a German writer who has talked much about this change in social character, that when young people are around 20 years old they don't have this fixed set of norms that we had a generation ago, because they are culturally liberated, and they have to start afresh when they are grown up; they have to build up themselves what we got from our parents. So it seems that the whole thing generates this situation. It might further the process.

Bohm: It could. You see, Prigogine has said that in certain situations a new order can develop out of chaos. What is required there, is not merely the equilibrium chaos of a certain thing at an equilibrium temperature, but rather a situation where there is chaos with a certain order in it. Namely, let's say, a temperature difference from one place to another. And in that temperature difference the chaos develops new forms of order. You see, we need some ordered field in which this chaos could develop, hence the new forms of order. I don't know what that would be, but as you say, the fact that the whole culture is being dissolved leaves people's minds not strongly set. But on the other hand, it gives no impetus there to do very much.

Q: But maybe even the terrifying threat from the bomb, the fear of all the chaos, might be the thing that could provide it.

Bohm: Well, I don't think fear can do it. You see, I think only love can do it. Creativity is not possible out of fear.

Q: No, but you see, although there has not been so much focussing on the Christian message of love, very many people have begun to realize that you have to develop this quality in yourself. And I think, actually, that the terrible threats we are living under are sort of making people realize that this is the way.

Bohm: Yes, well, I hope so. I mean, I don't know. You see, I'm suspicious of fear as a motive for anything because I think fear distorts.

Q: Since fear came up, can you tell me what you think fear is?

Bohm: What does it mean, right? That's the same question — what is it, and what does it mean, right? Well, first of all I think we'll begin at the primitive, animal level where there is a reaction of fear to a perceived danger, and this may often be a reflex. For example, I read somewhere that there were certain birds who have to be afraid of hawks, and when they take a hawk-like shape made of paper, the bird sort of freezes. That is, any shape of that form means danger, and danger means fear. That is, danger means that all the chemical messengers to alert the body must move out and make the body do whatever would be appropriate, right? Now that is the beginning of fear, where it makes sense.

Q: So fear is like a runaway loop with no block.

Bohm: Yes. The fear made sense at a certain level and still does. But the difficulty is this: that when the brain developed this large cortex which could then create out of thought all sorts of forms in the imagination, then this could also liberate fear. And that fear also works the same way; the fear disturbs the thought process. You see, thought does not work properly when the brain is disturbed by fear. All the chemistry is wrong.

Q: Maybe fear was very useful, as you say.

Bohm: For a primitive animal.

Q: Right. But now perhaps we have something else that is more useful than fear that we can put in its place — what we're developing — and that's that instantaneous knowing, or sensitivity to whatever. That inclination exists so that if there were a snake coming through the door, we could know it instantaneously before it gets here, so we can respond or act appropriately before it actually gets here.

Bohm: I think that is possible if you have not been heavily programmed about snakes. You see, I saw a program on television where someone was heavily programmed about, what was it? Spiders. And it took a terrific amount of work to get rid of that fear, you see. They were terrified of spiders. Now we have all sorts of fears, individual and collective. There is the fear of the stranger, the fear of the enemy. Now once fear takes hold, all the chemistry of the brain is altered. All sorts of chemicals change, the electrical distribution changes, the blood changes, and the thought process does not proceed with clarity. Now I think that our education often tends to foster fear, because a great deal of education is through fear. I mean, a child is threatened if he doesn't do right, and he's rewarded if he does. These are two sides of the same thing. That is, fear and pleasure and rage are really related. I saw some research into the structure of the brain. There are certain pleasure centres which when stimulated — they did it in a cat, and the cat looked very happy — and then by increasing the stimulation of those centres you could see the cat terrified, and with still further increase he went into a kind of rage which was pleasure again. And therefore every one of those emotions will affect the whole chemistry.

We have to face the nature of fear. You asked what was the meaning of fear, and what is its nature. So, fear is a reaction which started at the animal level and became entangled with the intellect and the imagination. And now, that fear disturbs the thought processes so that the greater it is, the more the imagination projects images that feed the fear.

Q: Can we go into what seems to be the opposite of fear? The opposite of fear is love. Would you see this as having a closeness or kinship to this flow of meaning, energy and matter?

Bohm: Yes. I mean it's very hard to say much about it, but the true creativity in this flow would only be possible with love.

Q: Perhaps love is that flow.

Bohm: Yes. It may be.

Q: Would you say something, David, about attention, because all the time you have been speaking, we have been moving our attention around into different areas, and does attention relate to meaning?

Bohm: Well, I think we have to give attention to the meaning. We give our attention to whether the thing makes sense, or doesn't make sense, for example. We give attention, first of all, to what is actually happening, and at higher levels we give attention to whether something is making sense or not.

Q: So it may be that by the movement of attention, which

117

we seem to have some ability to handle, we can allow meaning to have greater unfoldment, if you like, placing the attention on the meaning rather than the energy or the matter.

Bohm: Yes. And in order to do that, to some extent the action must be suspended, right? Now it's very hard to give attention to the meaning while you're actually doing it, though perhaps it is possible.

Q: Well, doesn't it happen at the point where, as you were saying, we are meaning?

Bohm: Yes.

Q: That's where the attention and the meaning become one — I am meaning.

Bohm: Yes. You see, what is attention? These words are very vague; their meanings are not clear. Now attention, according to the dictionary — its root is to stretch the mind, tension, at-tention, to stretch the mind toward something. That is it. In other words, you stretch the mind to come into contact with that something.

Q: What relation does attention have to intention, etymologically?

Bohm: Well, it's the same root. Intention means, I suppose, the tension within. The tension to do something, that state of tension out of which you act, right? Also the word 'tendency' is in there as well; the meaning of tendency, as the tendency within the intention...

Q: Attention seems to suggest the person alerts himself fully to what's on. He is fully with it. That is what attention is. We say for example, you ought to pay attention. Attention is giving yourself fully to what is on at any one time — giving yourself, in that sense.

Bohm: And that picks up the meaning, right?

Q: I was very interested in your soma-significant and signa-somatic relationship between body and the unfolding of deeper and deeper levels as you go on. Can you give us an example of an experience of your own?

Bohm: Well, I can't at this moment give an example.

Q: Would you like me to give one?

Bohm: If you like.

Q: You see, something happened in several levels, in working with a particular patient. This patient was given to meditation, the practice of meditation. In my work, I don't use meditation; it is simply standard psychoanalytic treatment of patients. At the end of a session, the third session or so after the patient started, she asked that when there was five minutes to spare, could we spend five minutes in meditation. I've never done that before. I myself meditate, but I'm inexperienced in this sort of thing. So the way she asked, I agreed. Now I had a question as to whether I should have done it, because my professional terms of reference would not be to mix the two things. So I left the question in my mind. Till then, my mental activity was with the psychoanalytic treatment of dealing with her material at the surface level perhaps, or the interaction level. After that, I closed my eyes as she did. I meditated; she meditated. And my meditation was, simply, to empty my mind, not with any thoughts, but just being. And it emerged, after I had closed my eyes, I suddenly saw, but with my eyes closed, a person rising about six feet away from me at my desk, symbolizing a soldier sort of figure. He walked forward, and there was a cushion in his hand. He went to the patient and put it at her feet. He walked backward, and then came forward towards me with a cushion and put it at my feet. My eyes were closed, and I immediately understood that this was

119

an answer from somewhere deep within myself that it was OK to let the meditation take place. I recognized him as a minor incarnation of the great Shiva. I did not know at that time what he did or why he came.

The next day, Saturday, I went to the library and looked up all about him. His mission was to restore the right relationship between the spiritual order and the secular order. There are more amazing things about this, but if we just say that in terms of these levels of unfoldment, first, I was concerned about leaving my ordinary, mental activity in my professional work, and second, I stood outside of it and asked a queston about the technique — mixing the technique — and there was no active, mental operation. Simply from somewhere within, this mental image just came up, possibly from some collective unconscious in which this image lay, and it just rose. From all of the literature that I know, this would be nearer to the answer to this question of how, in the body, some of these deeper mental activities are projected.

Peter Garrett: I think we are going to have to finish off soon. We have been going for nearly two and a quarter hours.

Bohm: Yes. It's been a long time. *(laughter)*

Q: It's been a marvellous one though.

MORE ON SOMA-SIGNIFICANCE
MEANING, SPACE, TIME, MATTER, AND MEMORY

Bohm: We ended up on soma-significance yesterday with this diagram *(triangle — page 91)*. Now I'm discussing its meaning. We have meaning, energy and matter. One can say that matter is a kind of condensation of energy that according to modern physics is interconvertible with energy. You can take matter, and it will disintegrate into energy. For example, an electron and a proton combine to give rise to light energy, or gamma ray energy; or vice versa, gamma rays can give rise to electrons and positrons. Now you can create matter out of energy or energy out of matter. They can turn into each other. And there is an interesting way of looking at that.

One view of matter is that we have a fundamental movement at the speed of light. The speed of light is, according to relativity, the same no matter what your speed is. You see, ordinarily you would think that the speed of sound is something you could catch up with and overtake in a supersonic airplane; but light cannot be overtaken. It's like the horizon, no matter how fast you go it is still the same distance away. That is the content of the theory of relativity. Space and time are relative to speed. We discover that we could ask the question: Suppose you moved with the light ray? As you move faster and faster and approach the speed of light, the amount of time between the begining and the end which is in your own frame, gets less, and so does the amount of space. At the speed of light, the beginning and the end of the ray are separated by neither time nor space. Not that we can actually reach that, but for the sake of imagination you can suppose it, and therefore you could say that something timeless seems to be involved. And spaceless — a fundamental relationship which is beyond time and space.

Light, time and space

We will diagram it like this:

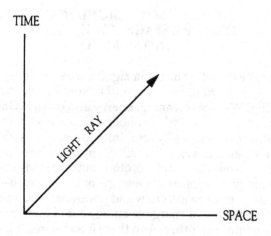

If you put two light rays together this way, they build up into time:

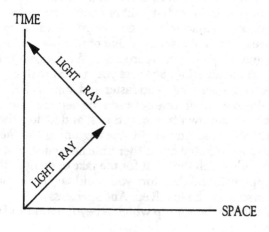

And if you put them together like this, they make space:

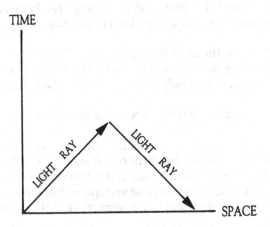

Eventually then you could put more in, to obtain structures of time and space out of structures of light rays, each of which are beyond time and space. That is, the fundamental interval would not be time and space, but time and space would emerge out of this — out of the structure of connection of these light rays, right?

Now current modern physics thinks of the electron as moving back and forth, approaching the speed of light, so its average speed is, whatever it will be less than the speed of light, right? Now in a way, the energy is being reflected. That is, instead of going straight on as with light, it gets reflected somehow back and forth. Now that reflection is what turns energy into matter. You see, the energy is condensed. And if the reflection ceases, the matter turns into energy. But it always was energy. So matter arises when there arises a pattern which reflects back and forth and becomes stable.

Q: What is it that is causing the reflection, then?

Bohm: Well, you see, we don't know. *(laughter)* That requires a deeper theory in physics, which we don't really have. But we could say that, in so far as there is reflection, then energy turns into matter. If the energy does not flow freely, it becomes a kind of matter; and that energy can be released by ceasing that reflection and making it go straight, right?

Q: And we say the same thing, don't we? 'What's the matter?' *(laughter)* When somebody is bouncing around, and it's not flowing freely, we say, 'What's the matter?'

Q: So matter is the interference of the flow of light energy?

Bohm: Well, that is one way of looking at it, yes — to say that matter is one development out of energy, which is light or anything similar which moves at the speed of light. Now therefore matter brings in time and space. But pure energy would not have time and space, according to that view.

Q: So that's why thought-forms deflect, to a great degree, the pure energy?

Bohm: Yes. Now we could say that it may be meaning that turns energy into matter, or matter into energy.

Q: You have used the image of energy as being a huge ocean with lots of swells, and matter like lots of little icebergs or lumps that congeal on the surface. I find that a lovely image because it enables me to get hold of a lot of things. But could you say something about where it's misleading?

Bohm: Well, the ocean is not actually a substance, you know, but it's an ocean of energy. When you come to empty space, this energy... It's hard to describe. That is, it's in the implicate order. It is not localized. And then you raise the energy a little bit above that of empty space — the energy of empty space is enormous — and make a ripple on top, then you get matter. You can raise it and get light, a form of wave which just goes straight on, or you can raise it in a way that makes reflection take place, and it'll become matter, right? Now matter is then a sort of ripple on top of the sea of energy which sort of involves reflection back and forth.

Q: Like a standing wave?

Bohm: Yes.

Q: Is that in empty space?

Bohm: Well, space may appear to be empty because matter and light, which are excitations of the vacuum, pass through it without deflection. So all we see is through matter and light. And light just goes straight through, as if it were empty.

Q: Would you say that matter and light are phenomena of nature, or are they things that only become manifest when one bit of nature, you might call it the sea of energy, interacts with another bit, which you might call the nervous system? In other words, do they exist independently of a perceiver?

Bohm: Well, that's a deep question. You see, I think that we are supposing, in some sense, that the universe exists independently of particular perceivers. The question you could raise is whether the universe perceives itself in some way through soma-significance.

Q: Someone said that matter is like thoughts crossing in space, and where they meet, it forms matter; and that sort of equates with this doesn't it?

Bohm: There may be a relationship. I merely brought that up to help explain the diagrams a little bit. But really, I was wondering whether people want to go on discussing what we said yesterday?

Q: You could say a few words on time. This word has come up many times, *(laughter)* and I haven't been able to really get a grasp on what time is.

Bohm: Yes, well, time is a very difficult subject. Let me, first of all, begin in ordinary experience. Now time has two aspects, you see. Man began to notice time as recurrence, the recurrence of the seasons, the recurrence of the days, the recurrence of the heartbeats, the process which recurs regularly and enables you to measure time, right? But then, there must also be something non-recurrent. Growth and decay, right? So you can say that certain things grow and decay during or after a certain number of the cycles of the recurrent process of the days and nights, or of the seasons of the year, right? So time, our concept of time, involves the interweaving of what is recurrent and what is non-recurrent. Now the suggestion is that time may be a way of thinking in abstraction from the whole process — from the order of process. This time order has a certain irreversibility in the sense that you cannot go back to what came before. It does not return, right? There is something recurrent, but there is also that which does not return, and nothing recurs perfectly, at least some things may seem to, like the atomic vibrations, but even then the state of the universe has changed since the beginning in such a way that even they have changed.

Now the question is about our experience of time. On the one hand, we experience time by being in synchronism with all this process going on in nature, and all around us. And we learn to think of it in terms of periods of time, breaking them up, and so on. Now time also contains duration — that is, these periods — certain things endure, while others change, right? Now if some things last a very long time, this would suggest, eventually, something which would be eternal, therefore beyond time, in the sense that it would cover all of time, right? Now also we can make durations that are shorter and shorter and eventually try to think of an instant of time. But everything that we actually experience has some duration, right? Now I think most religions have raised the idea that there is something beyond time. And this analogy in physics suggests that there may be something beyond time — that this light ray, in some way, is a structure that does not involve time in itself; but that time arises in relationship to these structures.

Now ordinarily we experience through time, and there are

many paradoxes in that because we think of past, present and future. But you see, the past is gone, and the future is not yet, and the present, taken as the dividing point between past and future, divides what doesn't exist from what doesn't exist. So that would suggest that there is no present. Another view though is to say that the actual moment of existence — this present — is in some sense beyond time. That is, everything that we see, and everything that we can recognize, depends on thought, on the past, on knowledge. This enters into our perception. We see all the forms and shapes, and so on. Without thought we would not be able to say what we see, or even see anything very definite.

But thought takes time, and thinking takes a lot of time, as I said. And thought, although it's very fast, still takes some time. Now ordinarily, when we look at this chair, for example, we say we can think about it and say we know it; but we only know its past. But for practical purposes, its present is not very different from its past. So that we really have a practical knowledge of this chair from its past, right? Now most of science and technology is based on that kind of knowledge; but when we come to processes that are extremely fast and subtle, then that kind of knowledge no longer holds. So if we try to look into our own consciousness and, say, look at what we are, what we will see is what we have been, right? The thought takes time. Therefore whatever thought presents us, the content of consciousness is a little behind the actual present. Now as I said, for the chair that's not important; but consciousness may — soma-significance may — reach into depths that are very subtle and fast. Therefore we may be missing the essential point.

Now many religions have raised the question of the timeless — whether there is not something beyond time — rather like this light ray, which would be the more elementary basis of experience or perception. Now time would be contained enfolded within this timeless state. For example, we can see already that present memory enfolds the past of the previous memory, and that enfolds the previous memory, and so on. There's a kind of order, like a set of Chinese boxes. So in any moment — in soma-significance — there is a kind of memory which gives the meaning of time, and this memory may not be entirely accurate. And of course, there's a kind of projection toward the future which has been even less accurate.

Q: Are you saying that there is definitely a direction, as in the nesting of the Chinese boxes?

Bohm: Yes. And then the next moment comes, and it enfolds the memory of this moment, right?

Q: That would bring out what we were looking at a little bit yesterday — the need to dwell in the present moment — talking about something timeless or beyond the dimension of time. To dwell in the present moment requires not dwellling in thought, because thought takes duration and is a slow process. We start to find a need, maybe, not to deny thought but to find the part of the experience which is before thought. Then thought moves on afterwards.

Bohm: Yes. That is really what is implied.

Q: How far can feeling be immediate?

Bohm: Well, in so far as it is affected by thought of course, it is not. Now there is a kind of perception which is immediate, but I think that the minute you distinguish it in some way — that is, through thought, right? — it must be a little behind.

Q: Is this suggesting that the transcendent state may be one thing that is now?

Bohm: Yes. And that now is also eternal in the sense that each moment — now — is always now, right? Now if you were to remove all the details, then there would be no fundamental difference of all the nows. On the other hand, all these details may be very important for other reasons.

Q: I didn't understand what you meant by each moment now is always now.

Bohm: Well, whenever we have this moment now, we have various details that thought provides which are in the past, but in the next moment that situation will prevail again, and so on.

Q: Could one have something here, like with music? You have various notes, some of which are falling away, and others are being sounded. You get patterns on those notes. You could have what is the present moment in fact, overlaid with thoughts and feeling which come later. And as long as you have a clear chord sounded by one of these, you say, 'I hit the mark.' In other words, they all correlate with a certain harmony. If they don't, the present moment overlaid with the thoughts and feeling that are coming through out of the past, don't give the feeling of hitting the mark. There's some relationship there.

Bohm: Yes. Our confusion is that we generally see the effect of thought fused with perception. The shape of things depends on the past thoughts and the way we react or respond to them. Therefore we do not see that that is the past. Now that past is important in a certain area. But in another area it might not be so important, or it might even get in the way.

Q: I was thinking that the problem is when you lose the sensation of the flow — I can't think of a better description. I mean that seeing the shape of things as being dependent on past thought is alright, as long as it doesn't get stuck — for instance, in judgement, or in assuming your assessments are the way things really are, and continuing to act on the basis of that assessment which was from the past, and so wasn't even accurate really when you saw it. So you need to just keep on moving without the past being the cause of your next action.

Bohm: Yes. One can say that psychologically the next moment need not be determined by the past, because it comes from meaning, right?

Q: Does all this amount to saying that the only direct

experience in the instant is the sensory experience, but that we use the word 'sensory' metaphorically? As Peter was saying, there's the thought in the past which makes the experience, but prior to thought there was the experience itself. Say for example, seeing — I see the thing; the seeing itself is immediate and direct; it's not in the past. It's the thought about what I have seen, and the memory of it that is in the past. And this applies to touching — all the senses. And this may apply to intuition, say, where you simply experience directly. It's direct in fact — prior to — and when it comes to awareness it's probably a question of illumination, or coming nearer to the light, as another metaphor. You are there, illumined, or your sensitivity is there, just to be picked up; you don't do anything more with it. That moment is present time.

Bohm: The problem is that thought works very fast. We have the program, and the program affects the seeing. Now the question is, can we see the program as the program, rather than seeing it in terms of its meaning? You see, the program has a meaning that is created mechanically, and that meaning is fused with the meaning of direct perception.

Q: I'm always coming across this statement in the literature which says something like, events exist prior to us in time. That being so, we cannot create anything, since all things exist already, and we can perhaps only discover. Does that mean that when it comes into the explicate order from the implicate, that is what you mean by creation?

Bohm: I think that if we said that we only discover things, that would say that the whole of time and space are already determined, and we merely find out what's there, right? And then we would raise the question: What about us? Who are we? Are we part of time and space? So are we determined too in that way? Now I think that it would be better to say that time and space are not determined, but rather, that they are like meaning, they are part of meaning. Now we said that meaning is ambiguous, and that leaves room for creativity.

New meanings emerge, and therefore something new emerges.

Q: Where do those new meaning emerge from?

Bohm: Yes. Now if we raise that question, you say we have meaning, energy and matter. Now they might emerge from energy, or from matter — that's one way of looking at it. But then energy and matter are also forms of meaning; so I think that we finally have to say that perhaps that is not a right question. It makes some assumption. You see, I'm only exploring now.

Q: It could also be an inherent aspect of meaning. In other words, meaning may inherently change in itself, and the programs may be a denial of that.

Bohm: Yes. You see, meaning has the character of being able to go into the meaning of meaning. Now where that comes from, I don't know. You see, at some stage when we are discussing something, we have to say this is our starting point, which we might question later.

Q: We are continually adding perception, and when we add a perception we expand the context, and as we expand the context, the meaning and the context will both change for us. And therefore it's a continuum in that sense.

Bohm: Yes. The change of meaning is also a change of being, right? Now one of the suggestions I made yesterday was that something similar to meaning may occur without us. So we could say that meaning is not just meaning for us, you know, but rather, meaning is already a dynamic activity.

Q: In these terms, do you have an explanation of a psychic who sees the future? How does that fit into this model?

Bohm: Well, I think that a psychic would not see the

future, but see what is a likely future. There are cases where they say they've seen the future, and the person behaved differently — he didn't take the airplane which was going to crash — and therefore the future was changed, right? So therefore a great deal of the future may be enfolded in the present as a potentiality, or a likely possibility, but still be changeable.

Q: I'd like you to talk about meaning — the meaning of meaning — a little bit. There was something I found quite interesting that happened in our small discussion group yesterday. We were talking about the meaning of meaning, and actually getting rather bogged down with fragmentary thought. Then someone suggested that we pause and have a few moments of silence, which we did, and what I would describe as a communion occured between us, and a resting of thought. Well, something else began to happen when someone spoke spontaneously. It was as if there was a rise in the energy level, and a clarity of perception arose, and some more true creative thinking began to happen. It seemed to me that this would have something to do with the energy that flows between people, of the nature of friendship one might call it, or some space to allow some intuition, some perception, to come into the picture to dissolve old programs perhaps. I wondered if there was more you could say about the meaning of meaning, and the relation of these points on your triangle — how they affect each other, and how they work together, and how they, in our relationships with each other, how they might be applicable.

Bohm: Well, as I said, meaning organizes energy and may organize energy into matter, and meaning may arouse energy, right? I think you have to say that these are like three different aspects of one whole, that they're merely abstracted in thought. When we are caught in thought then we have something like a kind of matter going on which has to be dissolved into energy. So seeing the meaning of that may help dissolve it.

Q: In our discussion yesterday evening we acknowledged that friendship had a place that somehow was beyond time, so that next time you meet, the friendship still exists. And that that had a part to play in the dialogue and with meaning. Friendship doesn't seem a very scientific term to be dealing with. I wonder if you could place that in a...*(laughter)*

Bohm: Well, it's hard to put that... I mean, if people are working together and have a sense of trust and get to know each other, something emerges in their whole relationship so that they can enter this friendship. It can come into activity when they meet, right? But it is always there as a kind of enfolded potential.

Q: Would you define energy for me, please?

Bohm: Well, you see, I think these are very basic terms, like meaning, that are very difficult to define. You see, in physics, energy is defined as the capacity to do work; but then, what is work? If you look in a dictionary you will always find words defined in terms of other words, and those in terms of other words, and eventually some of the original words will come back if you follow it, so that you don't really get a definition in the dictionary, right? But what happens is, as you explore and unfold these meanings, you begin to perceive what it means. Energy is this power, this movement, which acts, right? Now it has to be given a direction, and meaning does that — a form — but meaning may also arouse energy. You see, energy may be in a dormant state, or in this state of moving back and forth and being trapped, and a new meaning, the perception of a new meaning, may allow that energy to be liberated.

Q: That's just what I wanted, thank you.

Q: I just felt that we haven't really finished sufficiently, if you can finish, with the previous two questions of Anna

and Graham. Could we explore a little bit further into
that — that friendship or communion, as Anna said.
Could we go into how it leads on into this notion of
community, which I feel is very relevant.

Bohm: Well, based on a certain amount of trust in
working together, you establish a relationship in which
each person enfolds the other. There's a flow back and
forth. And now what's crucial for community is free
communication. That is, whatever happens has to be
communicated. Through communication people come to
one mind; that is, to one meaning, right? You see, if we
all have different meanings, if the situation means
something different to each of us, then we are not going to
be able to work together. So even with the bees, they have
a common meaning to the dance. So to establish a
community, we must have easy and free communication
without blocks. These blocks come from the program.

Q: One thing that we did bring out that could probably
bear further consideration, was the fact that we avoid the
experience of mental pain by not being in the present
moment. And that being in the present moment is the only
place in which this creative act of change can happen; it
can't happen in an imaginary past or an imaginary future.
But to avoid that, to avoid facing that, we kind of fudge
the issue a little bit. And we were linking this in with the
necessity for this safer atmosphere where there is
friendship, for this to be able to happen.

Bohm: Yes. We were discussing in one of the groups
yesterday that we find that certain questions are very
painful to face. Now thought is constantly trying to find
ways of escaping, of avoiding the question. Thought is
programmmed to try to do this and to conceal what it is
doing, right?

Q: Do you see it as possible to get out of those grooves,
and actually program one's thought to be a partner in

questioning relevant questions?

Bohm: Would it be possible to have this come from a program?

Q: Probably not. The program becomes increasingly irrelevant if there's a new meaning constantly emerging in each moment. I'm not even sure what the point of a program would be, if everyone shared that meaning.

Q: A self-destructive program.

Q: Well, it just wouldn't be needed, because there wouldn't be a past.

Q: But some programs, like knowing how to drive a car, or the sub-conscious program that keeps the heart beating...

Q: Right, right. There are some subconscious processes.

Q: We wouldn't want to throw that out.

Q: No. I mean, those are vital. It's just the ones that take over the role of conscious creativity, that conceal themselves as creativity or spontaneity, when in fact...

Q: How in fact do we know which is which?

Q: Well, this is why friendship is here.

Bohm: Now we were discussing last night that people talking together might discover that a person had some problem, a block which might be painful, right? Now in the presence of pain, thought is programmed to constantly bring in thoughts that would be more comforting or pleasant; but they may be illusory; they usually are. Now these usually concern thoughts of the future when there will be less pain, or else the past when there was less pain, or what might be. Therefore we were constantly slipping into the past or the future, but both were really the past, right? And therefore we were not looking at the actual present.

Q: I would have thought too, that blocking — this blocking process — may appear to disrupt progress, but may allow progress to be made. In other words, you have a recovery time given to you, perhaps where you recognize it, which enables the brain to sort out the thinking process. Probably the blockage is important, provided you are determined to solve the problem. I'm thinking of negotiations in industry, where you know there is something very serious going on and your task is to solve it. Provided you know that, and the blockages occur, and you know they are bound to occur, this could be a very good way that the brain has of giving you a recovery period.

Bohm: In some cases that could be so. But you see, many blocks continue indefinitely; we are not conscious of them. You see, in general there is an attempt to avoid consciousness of the block which is painful, right?

Q: Yes, but in real negotiations you have got to own that, otherwise you don't make progress. It may take a long time, but the mere fact that the problems are posed, in the end gives people a chance to find solutions.

Bohm: Yes. Well, I think if it comes to negotiations over actual problems that thought can handle, then that is alright. Now the question is whether thought can handle the kind of blocks that occurs in psychological problems.

Q: I would say in that situation, the lengthier time just indicates that there is more time being spent trying to avoid the blocks; the blocks try to hide themselves, and you're trying to get them out again. If you go straight to the heart of the problem then the block has no function, or the block is bypassed completely.

Bohm: How do you do that?

Q: I don't know. But I'm contesting the idea that a block can be useful as a resting period. Ideally, it should be just simply bypassed.

Bohm: Yes, well I think that if you are doing something practical, you may hold back to give time to sort things out. That's what the gentleman was saying.

Q: That's right. Yes. Because what we could say, David, was if somehow you've established a rapport, there is this kind of friendship which will protect you as you go along. And therefore hopefully the solutions will emerge in spite of the blockages. Thus in time you can go back to the blockage and approach it from a more intelligent angle.

Q: Isn't the problem here though, that the roots of the blockages are not at a level which can be discerned by conscious thought — that they are in subtler levels? And this is where the friendship that Graham brought out starts to penetrate into the implicate order where it's subtler than something that can be discerned and discussed by saying, 'Well, I think your block is around this area, and my block is here.' Because actually it's at a much subtler level that there may be some freeing of those blocks.

Bohm: Yes. I think that friendship can do that.

Q: This is important. That relationship is important.

Bohm: Now the question arises that sometimes you have to stand alone and deal with this.

Q: Would it be the truth, if one is having to stand alone, which I suppose ultimately we need to do, that the block, the awareness of a program, becomes clear when we bump into the wall, when reality kicks us back? That says, 'Ah, yes. Change is required.' So in a way, aren't we talking about the need for changes — the shifts — the changes in meaning which you say involve changes in being, or changes in being which involve changes in meaning? That in a way, there is a need for constant change in meaning, or constant availability to change? The image that keeps sticking in my mind is a clear image that fades into fuzziness around its edges, as opposed to the needs of a program which demand very precise frames within which to work. And I don't know quite where that's leading me, other than that it's an image of flow and flexibility, as it were, in one's state of consciousness.

Bohm: Yes. Well, what we were suggesting last night was that when it comes to some really deep problems, there is this continual tendency to escape. You see, there may be fear behind it, right? Now any intense pain or fear distorts the reaction of thought.

Q: You spoke about our wanting to liberate this desire to avoid the issue, and you suggested that in sticking with it and not doing whatever the program suggested we might do to avoid the given issue, that there might be a liberation, or a build-up of energy sufficient to penetrate to the lower, more implicate levels, and actually free up what was really there. And I just wanted to pull in what we were saying about communication, because we were talking about standing alone — that when something comes, I won't be deflected by my familiar thought patterns, and I will stand alone, and see what comes out of that. And sometimes a way of not standing alone is to communicate. And yet we were talking about communication and friendship.

Q: Standing alone lets something clarify...

Q: Or allows something to intensify through communication, which is actually what we are doing in

138

this moment. It obviously requires a certain willingness on the part of each one, so the pressure builds. And the extent that the pressure will build is in direct proportion to the willingness of each one to basically stay with it through whatever. And as we were saying last night, it takes honesty, and kind of old-fashioned values, right? And a bit of passion. Because one begins to see that the program isn't simply something that has affected us individually, in our own actions, but is basically affecting the whole. And to the extent that we allow ourselves to let this program that has been running us dissolve, we have an impact on the whole.

Q: I think those two things don't really conflict for me — that of standing alone, standing back, or the ability to get rid of a block by communicating with a friend or somebody. It's really useful to have both things available.

Bohm: Yes. I think that whether you are alone or with a friend, there has to be this willingness to face the pain and the fear, and to stay with it. Now this is where time comes in. The thoughts are constantly bringing in the past and the future, or what might have been, or what is possible, and we see that each time that happens, we are off the mark, right? Now if you keep on seeing that, then this generates the energy and passion needed to reach the deeper levels.

Q: So is that the force of necessity that you were talking about, where you get so fragile that you may fragment, or you get that desperate need of the creative urge to either rise above it or beyond it, or you just disintegrate? It's like one way or the other. Is that the force of necessity?

Bohm: Yes. Well you see, necessity is a meaning. It means it cannot be otherwise, right? It doesn't yield. Now the force of necessity is in the block. It doesn't yield. And there's a force of necessity in the meaning of understanding what's happening, which starts to dissolve it. Now you see, that's the power of meaning.

Q: So meaning might be something which creates a force of necessity, and the depth of meaning might be associated with the depth of necessity — a depth of unstoppable force to counter the lower depth of immovable object. *(laughter)*

Q: The question I have here is: At a denser level we might see a program to be a program, and it might cease to have the sort of meaning it had because of that, and so we may release it. But at a finer level, into something more implicate, we may have a pattern of friendship which may be effective in the same way — may create the climate for that change or release to occur. It may be that, rather than me seeing the patterns in myself, I might see the patterns in others, or they might see them in me, and it may simply be the understanding of them that releases them, rather than it having to be done individually within oneself. This brings out this interrelationship we were talking about yesterday, right through humanity. It may be sufficient simply to realize the patterns somewhere in the consciousness of humanity, rather than each person working through their own complexities, if you know what I mean.

Bohm: Yes. Well, that's an interesting approach. If it's realized with sufficient energy and passion then it may be able to work on somebody else, and spread.

Q: It's the technique of healing, isn't it?

Bohm: Is it?

Q: In healing, one person attempts to help another, so it must come in the pattern that was just described, I should imagine.

Q: If the holographic analogy is accurate, then each part contains some of the whole, and so presumably whatever takes place in myself — a clarification — has an effect on the whole. And a corollary of that is that often I find in working in interaction with the world in which I dwell, I often can't tell if it's that person over there who is a pain, or something in me that is a pain. And I really don't know sometimes which it is. And all I can do is to allow the change in here. I don't really know about that person, so I don't know that there's any difference between them and me.

Bohm: Yes. But not fundamentally though.

Q: As a whole, in the system there is pain, and who it is, one doesn't know.

Q: That's where one can get to work.

Bohm: It's like asking is the pain in the finger or in the brain?

Q: You mentioned a word a little while ago, David, which was 'trust' which seems to me, whatever it may precisely mean, is a potent word here, in industrial disputes, in blockages. If there isn't trust, change seems to me to be almost impossible. Sometimes you find with a social-worker, when a social-worker is kind of generating care all around her, but she doesn't, or he doesn't somehow manage to generate trust, that whatever he or she has to offer is rejected because there is no trust. And on the political front, the coal-mining position at the moment, nothing can ever resolve it because there's no trust on either side. So I think this word 'trust' is a very important one in this matter of friendship, relationship, changing, accepting one another's patterns, the whole thing. Nothing really happens unless there is this curious relationship that we call trust.

Q: Non-trust is an energy as well, but it doesn't work very well.

Q: Destructive energy.

Bohm: Yes.

Q: When you've got trust, you see, you've got to give trust a framework to work in. If you find yourself with strangers, there are tactical things that go on between the people concerned, and if you've got a format that you know is going to work, or usually does work, then you've got a start. Very frequently we found that in the midst of all the turmoil, the social life was important, the going together to tea afterwards, the sitting down and really getting together as people outside the framework of the previous discussion, informally over the tea table, you know, high tea or a bottle of beer, and just talking, so that they understood you and you understood them. It's another step toward this understanding, this wholeness, based on trust and friendship, and in the end you got there.

Q: I would like to bring in another program at this stage, which is the genetic program. And what excites me over the past ten years or so — what we have learned about the genetic program — is that there are lots of rearrangements going on even within the individual lifetime, within this program. So I think that comes in in connection with changing thought programs and changing even our genes in the lifetime of one individual.

Bohm: Yes. Well, I think there are two questions. This question of trust should be addressed, and then we'll come to this other one. Now I think this question of trust is a question also of meaning. This is seeing the meaning of the other person as basically the same, that we all have the same deep intention. And this new meaning begins to affect the whole structure, and begins to change it. It changes the thought process.

Now about the genetic thing: You've called attention to the fact that something similar may be going on. You are saying that the genetic process changes even within the lifetime of an individual, right? Genes actually change, whereas before it was said that they changed only at random, right? Now therefore the whole way of life may affect the genetic structure. In fact, you could say that thoughts are like the genes of meaning. You see, the genes are said to contain information, but information without meaning is pointless. Now meaning depends on context, the meaning of the gene's information, and which proteins are around it, and so on. But the genes may have indefinite potential for meaning according to the context. So not only does a rearranged gene have a different meaning in the same context, but in a change of context the same gene has a different meaning. Now that would suggest something about evolution, that something analogous to meaning is going on at the biological level. I developed this further in the part of the paper that I didn't give, that in terms of the computer one could get a similar notion that the information in the computer has a meaning in terms of the activities which flow out of it, or in the way it would sort out information coming in. If that's so, then even our biological evolution is going to be affected by what we do, right?

Q: Is it illusory to imagine that when all the programs have been dissolved, there is underneath this, what is called the essence of our being? Or is it that deep, deep down we are very shallow? *(laughter)*

Bohm: I think that in the next hour we are going to discuss the ego, but you are raising the question which also is the question raised by religion — What is the essential nature of being, right? And is the ego, or what we usually call the ego, very shallow, or is it something deeper? Or are we part of this whole?

Q: When we talk about dissolving the programs, there are just more programs to keep dissolving.

Bohm: But also they are continually being made.

Q: Perhaps we pay too much attention to them, and we're giving them power; we're putting the power in. The better thing would be the aspirations, the inspiration, it's perhaps not manifest within us, but if we hold to that, that will be the better life, rather than keep on empowering our blocks. We don't necessarily always dissolve them. But I mean there are more way of dealing with a brick wall than knocking it down.

Q: I would like to suggest that the intentionality to get rid of the block is crucial. And once we have committed ourself to that intent, then we can use vehicles that are close to us or that we have some familiarity with. For instance, we have all kept talking about the recognition of a higher intelligence within us, that is part of us. And many people consider that our minds continue to work at night, or while we're busy with tasks, so it's not far-fetched to ask ourselves about our blocks, and ask ourselves to dream about our blocks. Because dreaming frequently comes in symbols we can tolerate, that have many levels, and in the process if you can remember your dream and write it down literally so that you don't get into any judgement about your dream. At that point a trusting friendship is really helpful because a trusting friend can go over the information with you and stand off from that information and be able to see the context in a way that you may be still fearful of. And that process can go to various levels, but you begin to get the skeins of information starting to come through to you, of what your block really involves.

Bohm: Well, I think that's very valid, that as you say, the dream is put in terms of images or symbols, and in the dream these are less alarming than to state in words what they mean. Now the question arises: What does the dream mean? Now to see the meaning of the dream would then constitute a change in the block. I think there are various

ways of trying to approach these blocks. They may take very
many subtle forms, and we can't give any final version of
how to do it.

Q: No. Not on any of these problems.

Q: One small point: There seems to be this view that it takes
two to trust. What about the idea of trusting oneself?

Bohm: Yes. I think it is necessary to trust yourself. And out of
this can come the trust in other people.

Q: I think that regarding trust and friendship, that it exists at
many levels, and our understanding of it is probably limited.
I think we have all experienced friendship which is actually
part of the block — like having cronies. I think that there are
levels other than that, which I haven't experienced yet. But
it's an article of faith that they exist, as it's an article of faith
that there are shades of meaning beyond the meanings I
already know, and more meaning which I cannot find but
that I inherently know something about. I suppose it must lie
as a belief that friendship lies at very deep levels and is
accessible, and that this is the way ahead. It's more precise
than what we currently know, yet there is something there
that is perhaps going to work. But my concern is to know, to
be available myself to those deeper levels, whatever they
might be.

Peter Garrett: It's close to eleven o'clock, David, to remind
you.

Bohm: Yes. Well, I think that if you have a feeling of
complete openness between each other then the meaning can
flow, and the whole thing can work. Now I think that would
be spontaneous generally starting with young children; but
we have had a lot of experiences through life which have led
to these blocks — these programs — which get in the way.
The question is: Can friendship begin to dissolve even those
blocks?

Exploration

Q: I'd suggest that something has been occurring in this room where I felt a kind of struggle initially for comprehension, and then an increasing willingness to move from our habitual linear way of getting things ordered to letting a number of almost random thoughts unfold together to find out what happened, and for me anyway, that takes a lot of trust. I rather like to appear that I know what I'm talking about. *(laughter)*

Bohm: We all do, right? *(more laughter)* Yes. I think we must acknowledge that we are in a domain where we don't really know what we are talking about. We're exploring, right?

Q: Yes.

RELIGION, WHOLENESS AND THE PROBLEM OF FRAGMENTATION

Bohm: Welcome. I originally planned to give a talk entitled 'Religion as Wholeness and the Problem of Fragmentation' which I gave originally at St. James' Church in Piccadilly, London; but when I looked at it again this morning I realized that my views had been changing, and it was partly as a result of our discussions together. I hadn't really fully understood the implications of my soma-significance paper in connection with this paper. It's rather like Newton — people hadn't understood the significance of all the evidence showing that the heavenly motions are not different from the earthly motions — and suddenly Newton saw it. I don't think we can compare this with that, *(laughter)* but I began to wonder. It seemed a certain part of it — on the ego — would still be worth discussing. The idea occurred to me, so I went into Peter and Don when we met this morning, and we discussed it and they suggested the same thing, that they themselves felt that the latter part was more interesting than the part on religion.

I think this shows how friendship is working; that we have similar feelings and we are communicating them to each other in this situation. Another thing I would say is that, especially in this last hour, I'm not trying to be here as your teacher or as your authority but, rather, I too have blocks and fears, and so on. Therefore we are all together in this, and the fact that I'm on this elevated platform doesn't mean anything.

I think we were discussing earlier the problem of fragmentation. That's one of the things in this paper. And we discussed the general fragmentation which has been going on in mankind, and building up. Now it had many causes that we went into. One of them was science, which is taking a fragmentary form and helping to produce fragmentation. The basic cause there, being that the content of thought took itself to be in correspondence with a reality other than thought. Therefore the divisions in thought were taken to mean objective divisions in external reality, or in reality other than thought, and with that meaning, we would act accordingly and start to break things up into things like nations and groups, and in all sorts of ways.

Now science enouraged that greatly in its mechanistic development; and now religion, although its aim is wholeness. The basic feeling of religion is the yearning to wholeness. The very word 'religion' is based on *religare*, meaning to bind up, or it could be *religere*, which would mean to gather together, and the word 'holy' means whole, and so on. There is man's urge toward wholeness which expressed itself both in religion and in science. If you wanted to get a whole view of the universe, you could do it through science and philosophy, and also a whole view of being, through religion and philosophy. The East specialized in religion and philosophy, and the West was more science and philosophy.

Nevertheless, religions became a major source of fragmentation. The reason for that is very simple, because you see, in the beginning — I'm not sure if all the primitive religions would have agreed, but more modern religions have agreed — that the ground of all being is somehow permeated with a supreme intelligence that is creative, and evidence for that is the tremendous order in the universe and in ourselves and the brain. And then, probably with less evidence, but it's there, this ground is permeated with love and compassion, although sometimes it doesn't seem so, seeing all the suffering in the world and the destruction that goes on. Now I think that if people would stop there, then they would say God has that nature, or the mystics would say that the Godhead, the ground of God, has that nature. Now if we would stop there, probably people who have religious feelings would not disagree.

But now to go further, and say that God or the Godhead has such and such a nature creates divisions because one person, or one group, says one thing, and another group says another. These divisions are very serious, particularly serious because we are dealing with the nature of the absolute. Therefore there is no way to bridge these divisions, you see. It's either so or not so. Therefore when it comes to religion — religious differences — it must result in extreme fragmentation.

Now that was some of the content of what I was going to say. But you can raise the question now about this ground in view of soma-significance, and so on, and see if there is an ultimate ground, and we could discuss that. Religion, in its modern form, and probably even in the earlier forms, does presuppose this ultimate ground of being out of which all emerges or unfolds.

Now although science and religion are self-world-views, they have been important sources of fragmentation; but I don't think that they are the major ones. Something much more powerful and pervasive is the identification of self or ego as absolutely separate and distinct from others. What is relevant here is not only the individual ego, but also the collective ego in the form of the family, the profession, the nation, the political, religious ideology, and so on.

When I talk about the ego, I want to say that the ego is not necessarily the whole of the self. The ego is a certain feature of the self which we could call the self-image. In the way we have been discussing it, it gets taken for the self. The word 'self' has as one of its deep meanings, the quintessence, the thing itself. Fundamentally all human conflicts arise out of the attempt to protect such ego interests, which are generally regarded as supreme, over-riding everything else, and not open to discussion or rational criticism. Indeed, even the fragmentation due to the scientific and religious self-world-views, arises ultimately, because the ego, individuallly or collectively, takes these views as the secure basis for absolute certain knowledge about itself. We could ask why people, in this absolutely unyielding way, would insist on their religious views, unless somehow it has to do with the security of the ego. On the other hand, perhaps these religious views contribute to the being — to what the ego is at the time.

Now it has been the aim, tacit or explicit, of all religions to change the ego so as to end fragmentation. Ideally, this would be accomplished, for example, by healing the 'sin-sick soul', but failing this, there is an attempt, generally at least, to control the ego and limit its destructive effects by various moralities and other restrictions. Science in the form of psychology is also trying to accomplish similar ends with psychotherapeutic techniques and with drugs produced by modern chemical technology, and various other ways. But it's clear that neither of these ways has gotten very far — even in such an evident and obvious problem as the possibility of the extinction of mankind that is implicit in our general fragmentary way of life.

It seems to me that neither science nor religion has produced an answer. So it seems important then to inquire more deeply into why the ego is such a hard nut to crack, you see. *(laughter)* And in such an inquiry several questions arise immediately: Why is the ego, individual or collective, so important? Why must it be considered to be essentially correct, and always right? Why do people explode into violence and anger when they are insulted personally, or even more, when family, religion, nation, or ideology are treated in what they regard as an outrageous way?

Well, to discuss this adequately will require more time. Perhaps we can go into it. But I would like to suggest something relevant, drawn from the story of Moses in the Old Testament. In doing this I don't want to discuss theology, or discuss any knowledge, scientific or otherwise. Rather I hope that what I say about Moses can be regarded as nothing more than a story that helps give rise to a certain kind of insight into the ego.

As you may recall, Moses spoke with a voice in the burning bush. You see, when the voice spoke to Moses, it said, 'I am that I am,' and when Moses asked who he should say sent him unto the children of Israel, the voice said, 'You shall say, "I am", sent you.' From this it is clear that the voice was saying 'I am' is the name of God. Now you can see what this means by considering that in Moses' time there was still a fairly strong survival of man's primitive animism — a tendency to see everything as a manifestation of a living soul or spirit — and this view implies that all life is one. The spirit of each thing is enfolded in that of the other, as each person enfolds something of the spirit of others in his consciousness. So in a way, everything would call itself 'I am' if it could talk. *(laughter)*

Now a particular thing is characterized by attributing predicates to 'I am'. Qualities: I am here, I am a human being, I am strong, weak, rich, poor, and so on. If no predicate is attributed to 'I am', this can only mean the universal spirit creating and underlying everything. And this is also most deeply what is meant by the word 'God'. So you could say that the insight of Moses is that 'I am' is the natural name of God, or the Godhead, because of its meaning. It points to whatever is meant. Now this implies that no image can be made of the universal 'I am', as this would attribute predicates, you see. The ancient Hebrews, in fact, had strong injunctions against such images, and went further, saying that even the name of God was too sacred to be used except possibly for the most holy of purposes. But unfortunately as time went on, they attached to this name a vast range of attributes such as great, wonderful, magnificent, powerful, merciful. They began to characterize Him verbally, creating verbally-based images.

Now why is it dangerous to attribute specific qualities to 'I am'? This is because 'I am' without predicates, already means the universal intelligence and energy on which all depends. That is, the meaning, energy, intelligence, all the things that we've been talking about. That is, the name of this triangle is 'I am'.

Q: It's intrinsic.

Bohm: Well, we don't want to start on theology. *(laughter)* It's something that I just saw at this moment.

This energy — everything — depends on it; it sweeps all
before it. Now if you attribute this energy to any kind of
predicate, you put a limit on what the original 'I am'
means; it limits the meaning. The meaning, remember, is
power; it is energy. We're putting a limit on it, but
implicitly it still means the whole energy at the same time.
And so we're trying to pour the whole energy into these
limits. It must give this limited thing tremendous
significance and power.

Now this is a contradicton. It bears on what is of
supreme significance, and will have a powerful, disruptive
effect on the mind — hence, on the brain. Disorder, and
eventually subtle brain damage may result from this. You
see the power of these meanings. The word 'God' is a
rather arbitrarily chosen word. It might as well have been
something else; it could have been turned around the other
way. *(laughter)* Well, 'dog' could have meant 'God', and
'God' could have meant 'dog'. They are arbitrary symbols,
right? You see, the name, 'I am', is the natural name of
God because of its meaning.

Now the natural name of whatever you would mean by
whatever is fundamental. That sort of gets us across this
little bit — this notion — that we are not looking at
something other than ourselves on this diagram — we are
part of it.

Now, as I said, it's not in religion or in science that this
disorder is most acute, but it's in man's attempt to identify
himself. He does this by saying, 'I am X,' whatever X is.
But as I stated earlier, whether he likes it or not, 'I am'
signifies the universal energy, and X signifies something
particular and limited. One can see the essence of egotism,
individual and collective — to give the significance of the
unlimited to the limited — and therefore you say 'number
one comes first,' and so on. There are other sources of
egotism, and so on, other ways of explaining it; but this
might be regarded as a very fundamental way of looking
at it.

It often seemed very puzzling why the immediate
interests of the ego so often seem to over-ride everything,

even things people indirectly regard as very precious, such as life, love, happiness, and so on. The explanation implicit here is that we generally behave as if the ego regarded itself as the universal 'I am' beyond all limits of time, space and conditions. For example, if the honour of your nation is attacked, you may be ready to respond ultimately with nuclear bombs, risking the destruction of the planet and even the nation itself, but somehow what is eternally right will have been vindicated. In other words, this eternal rightness will still prevail, even though everything else is gone. *(laughter)* I think that's implicit in the thought, right?

Now this behaviour is implied to be absolutely necessary when any particular predicate is identified with 'I am'. The meaning of absolute necessity is to sweep all before it. 'Necessity' means don't yield; it means the object that doesn't yield, and it also means the force that sweeps all before it. So it either leads to a blockage, a rigid blockage, or to a force driving everything in front of it. Now it's not easy to change all this because it's deeply imprinted in the brains of four thousand five hundred million human beings. Merely to exhort people to think otherwise would have very little meaning. To do this would just be to superimpose a contradictory meaning on one that is ancient, subtle, pervasive and deeply held. So you would now have several meanings fighting each other; and this means that the most powerful was going to prevail. This would therefore tend to increase fragmentary egotism rather than ending it.

The challenge is to dissolve this old pattern of thought and perception rather than to try to contradict it, to control it, or to destroy it by force, or by will. Now again we could say that dissolving this pattern is the healing of whatever the real self is. One must say that there is a self — an individual self — but it's not the whole thing, and it has to be seen properly within its limits, although it may be far greater than we know. But still, it must be limited.

Q: I don't follow that.

Bohm: Well you see, our real self is nothing but 'I am'. Or if we have an individual self and that is real, then that must be limited to the extent that it is real. You see, this is an unknown question.

Q: Limited but not bounded, would you say?

Bohm: Well, there are no definite bounds to it. It could always, you know, unfold. And it has an infinite potential, but in some way this is still not the total.

Q: But vortex-like?

Bohm: Yes. It's enfolded in the whole perhaps.

Q: So there are two things in parallel there — two things together. One is the self — the level of 'I am' — and then...finding words is quite hard here...and in the context of that, there is a self which isn't unbounded. And when one misses that point, one applies infinite...

Bohm: It may not have any boundaries but in some way it is limited, in the sense that it isn't the whole. Like the surface of the earth has no boundaries.

Q: It's dimensionally less localized.

Q: It's unbounded.

Bohm: Yes. It has unbounded, infinite potential, but it's a different order of infinity from the whole.

Q: But if it is reflective of the macrocosm and the microcosm, then it would not be limited the way you suggest.

Bohm: It may enfold the whole. But you see, it's like the hologram, which enfolds the whole but in an incomplete way, right? So still, the danger is to identify this with the

154

total 'I am' — the total meaning of the word 'I am'.

Q: Which is something very empty.

Bohm: Yes, but it may also be very full. You see, when we apply a predicate to 'I am', it limits it, right? Now in some way this thing is not limited; even this description will in some way limit it probably. Now you could say a little bit more. You see, this word 'I am' has an interesting character in Hebrew, or in the Aramaic language which has the same grammar and was used by Christ. In Hebrew it is not possible, really, to say 'I am' in any proper way. That is, you could say 'I here', 'this table.' The present tense of the verb 'to be' is not used, right? Now in fact, in order to say this, the voice in the burning bush said *'ehyeh'*, which means 'I will be whatever I will be', or 'that I will be that I will be'. So 'I will be' was taken as the name, right? But it was translated into English as 'I am'.

Now you can look at some of the statements attributed to Christ, like 'I am the way, the truth, and the life.' Now it's hard to see how they could be stated within that grammar. You see, if he had been stating it personally he would have had to say, 'I the way, the truth, the life.' Now if he had said 'I am', he would have had to say, *ehyeh*, and he would have been using the name of God, right? There is no verb 'is'. So that could equally well mean 'God is the way, the truth, and the life'. And similarly 'Before Adam was, I am,' doesn't seem to make sense as a personal individual, but 'Before all worlds *ehyeh*, which means God, is.' 'Before Adam was, God is.' The point is, that was translated into Greek as 'I am', a mistranslation, probably. It could be. The same as we discussed the mistranslation that occurred from Greek to Latin of *metanoia*, repentance, and *hamartia*, which is missing the mark, sin. That illustrates the responsibility we have in doing a simple job even, like translation. That can have very great consequences in some cases — to mistranslate.

The fruit of the tree of knowledge

Q: When we talk about 'I am', I am self, the level of 'I am', or 'the I am', this has brought us back to our being on the mark or off the mark; and I have an experience of there being a self which is on the mark or in that direction, and there being something which I identify as the possiblility of myself, which I am filling out ongoingly, and it seems to me... What you are saying... It's on the edge for me.

Bohm: Yes. These two things come together in a sense that when we are on the mark then we are taking part in this whole thing, and when we're off the mark, we're introducing some confusion in it. It seems that being on the mark about 'I am' is very important in some symbolic way, you see. That is where the trouble, perhaps of egotism, starts — that we go off the mark with regard to 'I am' — we don't give it attention.

The other day we were discussing the legend of Adam and Eve and saying that Adam ate of the fruit of the tree of knowledge, which included the knowledge of good and evil; and we were saying that something must be wrong with that. Out of knowledge came all that technical knowledge, and so on, but it included the knowledge of good and evil. Now we're saying that evil is merely the result of being off the mark. Therefore any distinction of good and evil is itself off the mark, right? And eating of the tree of knowledge which made such distinctions possible — applied to good and evil — was possibly the meaning of that legend.

Q: Distinguishing without judgement doesn't necessarily mean fragmentation.

Bohm: No. But in this there may be an irrelevant or wrong distinction, you see. It would say that good and evil are opposites and therefore related. Now, are being on the mark and off the mark opposites?

Q: One asks for an answer which is either yes or no. So

maybe one doesn't want an answer to that question.

Bohm: No. But there is something wrong with the notion that they are opposites. You see, opposites imply each other, like hot and cold, East and West, and so on. Now being off the mark is not the opposite of being on the mark. This way of putting it with evil is somehow wrong structure, or wrong meaning.

Q: It's more like a direction, or a course correction.

Bohm: Yes. You see, we would rather say, not that there are two things, good and evil, but rather there is the question of attention which keeps you on the mark, or failure of attention which makes you go off. The failure of attention is not the opposite of attention but it's due to the program. Does that seem clear? It's not intrinsically related to attention but it's entirely different. It's not like East is related to West. Now it's clearly a question here of meaning. You see, at this level wrong meanings can have tremendous consequences because meaning arouses the energy, or it contains the energy, gives it form and shape, and so on. And this wrong meaning is a very big factor in building, in making possible, this egotism. Now we can ask what can be done with this egotism. We have made various suggestions here, such as friendship will help to dissolve some of it, or you can look at some of the blocks. The major block is looking to the past and the future. You see, one of the points is that this structure not only blocks but it conceals its origin by moving to something else like the past and the future, or what may be, or might be.

Q: We have something in our language which might describe this. When one behaves badly, one says one forgets oneself; I forget who I am. And I think that perhaps one of the blocks is forgetfulness of who I am.

Bohm: Yes. That's important. I forget *who* I am, and also forgetting '*I am*'. *(laughter)* If you forget who you are then

I think you're mistaking yourself for 'I am'. But I am this
person who is a limited being, insofar as I'm being here;
although that may be infinite. But by forgetting
that, we're tacitly giving a value of unlimited.

Q: The feeling that I have is that any predication of 'I am'
is going to be false.

Bohm: Say, even love and intelligence, right?

Q: Right. So one has in 'I am' the indication of meaning as
something which is implicate, and as soon as you predicate
it, you've made it explicate, and it's false. So we have that
problem at a very coarse level when we say, 'I am a bank
manager'. When I'm not a bank manager anymore, I fade
out, or whatever. *(laughter)* We start to attribute qualities
beyond the natural sense of them — absolute qualities —
things which are unfolding and enfolding, which are
continually changing, and therefore we get these programs
which try to continue inappropriately.

Bohm: Yes.

Q: Why is that false, rather than incomplete? If I am a bank
manager, that's not all that I am; but at least it's part of what
I am, or it's part of the way I'm acting, let's say. Why is that
false?

Q: Only because the term 'I am' signifies an everlasting
absoluteness of the entire implicate which is beyond
conception. And to identify that as a bank
manager...*(laughter)*

Bohm: One of the things we fail to notice — or forget —
which the story of Moses implies, is that 'I am' already has a
meaning which is deep and pervasive, and we can't control
it. And therefore when you add 'bank manager', then that
meaning of 'I am' is attached to 'bank manager'. You see, it
gives power; it says the same as to say God is a bank

manager. *(laughter)*

Q: It becomes reasonable if we keep repeating it, of course.

Q: There is also a point — 'I am not a bank manager', is also equally wrong.

Bohm: Well, you could say 'God is not a bank manager,' which is sort of irrelevant. God is not chocolate, He is not cheese, and so on. *(laughter)*

Q: You know, in Italy there is the Bank of the Holy Spirit.

Bohm: Yes. I once made a joke to a fellow that you put your money in down here, and you draw it out up there. *(laughter)* He was a fairly devout Christian, and it took him some time to appreciate it.

Q: Nevertheless, in trying to raise some understanding of our relationship with God and some understanding of God, can it be useful to try and get some ideas of God within our own frame of consciousness and our own frame of reference, so that perhaps we can get a perspective of energy, or whatever, recognizing that it's terribly limited, but telling us something about God, and perhaps more than we knew before?

Bohm: Well, that's just the question we're raising. I don't want to give an answer to that because I don't know. But you see, we're exploring this question. Now it was implicit in what Moses said that any qualities attributed to 'I am' would be superfluous, right? Although they did do it. And it's also implicit in the notion of meaning. You see, the question is: Can we do this consistently, and say God is this, that or the other? Peter has just objected to my proposal that God is permeated with intelligence and love. Do you want to come in on that?

The problem of description

Q: Saying He's permeated with...

Bohm: Or He *is* that, whatever you like.

Q: ...is different from saying, He is that.

Bohm: I'm trying to get the language so as to avoid this problem, but still let's ask, is there any objection to that?

Q: This is not quite an answer to that question, but it's an offshoot of it. It occurs to me that in that ground of being, permeated with these attributes — at this stage I'm not even addressing whether they're predicated or not — we may experience love and compassion and intelligence, but when we think about them we think about our thoughts of what love and compassion and intelligence are, and we expect that all-pervasive love and compassion and intelligence to look like my personal picture of what they are.

Bohm: Yes. That's one of the problems, I think, of giving attributes to the unlimited.

Q: You see, if your proposal that God is intelligence and love somehow implies that therefore He will act this way, that way and the other way then, I think that's a mistake.

Bohm: Yes. But if it implies nothing then it's not really doing anything. *(laughter)* It's a red herring. It's confusing the issue to give names as if you are implying something, and it turns out that you are implying nothing.

Q: And yet friendship implies nothing and opens things up.

Bohm: Yes. Alright. But then that implies something — it implies that these qualities will help open things up.

Q: But if you look at God in terms of love — being pure love, or complete love — and we are less than pure love, and less than complete love, then we can get some sort of notions of relationships. And perhaps even behaviour.

Bohm: Yes. I would like people to answer that, because I don't want to make this just between the two of us.

Q: Haven't you got to do that? Because you can't just not say anything about God. Otherwise, all you're saying is, God is 'I am' and 'I am' is God, and you haven't said anything.

Q: You always seem to be attributing God as a person, separately. The Christ message is that He is the totality — all its manifestations and infinite variety of phenomena that we keep talking about. But this trying to pin it down is the wrong question; it's just the acceptance, and we are part of it.

Q: There's a paradox, or a tautology, I suppose. We say God is love, and what we mean by love is what we mean by God, and so actually we're not saying anything. So realizing that, we then feel compelled to start predicating this idea of whatever the ultimate ground of being is, and in the process losing it, and that seems to lead us to the point of... well, relying on, or looking to, the authority of personal experience as the only way out, which then starts equating personal experience with all that is, and can't help but cause a kind of fragmentation because it doesn't seem to be in the realm of available thought or anything else to grasp on to.

Q: It just occured to me that we've talked together a lot about using these words to predicate — or as attributes — but we've talked about words being used as metaphor.

Bohm: Yes.

Q: But perhaps that would be another way of looking at it.

Bohm: But then every metaphor is limited in some way, right?

Q: Yes, I agree with that and I think what we are doing is looking for language. We have really not a very good language for metaphysics here. And I think perhaps we've two different things we could be looking at — the structure of the universe in, I suppose, scientific terms, and then I think perhaps what I'm doing right now is looking at the meaning of the meaning. In other words, the language that I would use when I'm trying to find my meaning in whatever I do is saying 'Right, God is a useful meaning for me, a useful concept,' and I put that in terms of love as being the wholeness — a word for the wholeness. So when I get into these states of confusion or depression or whatever, instead of thinking that the depression, the unhappiness, the fear, is actually all there is — and it often seems like that in a fragmented moment — I can then trust that my meaning in life is what I would term love; and that sooner or later that meaning would come through and I would haul myself out of the depression, or whatever it was. That anchor of God, the meaning of that, would be love, and that would be the language I'd be using.

Q: I see a danger. It's not an absolute necessity, but I do see danger in the idea of God being love. Then my limited concept of love associates it with such things as happiness, joy, no pain, this kind of thing. Now I run a danger of going straight back into duality by saying 'Well look, there are places and experiences which are painful, which are torturous, which are hurtful, therefore there is evil as well as God.'

Q: Going back to the idea of language that John was putting forward, suppose hypothetically that you have two people who have had some experience in direct intuition of the supreme spirit, the supreme name of God; so they're talking and one says, 'Ah, it's beautiful.' The other says,

'Yeah, it's beautiful.' Now somebody else hears them speaking who hasn't had that experience, and he'll say they're limiting God; but actually because they both have a common experience, they trust each other. They have to express because it's natural. If you have a very profound experience or realization you have to express it; and here you have two people who have a mutual trust that there is a real experience behind the expression, and they know that they're each only giving a minute fragment, but still it's a meaningful exchange. But to somebody who doesn't have that experience it's going to seem completely inadequate.

Bohm: Yes. That's where the trouble is. The communication is limited to people who share certain experiences. If you want to establish and bring about wholeness, we must communicate in some other way. You see, the community is communication, and we must be accurate in our communication, and not miss the mark, right? This missing the mark in the translation of the Bible has probably introduced tremendous confusion and error and destructiveness in our lives.

Q: Excuse me. You are also allowing yourself some latitude. For example, you say we don't want to talk about meaning; we're going to let it unfold; so similarly, if we are talking about the total qualities of the whole, we could equally say let's not try to define precisely but let it unfold.

Bohm: Yes, but when we want to talk about it we have got to say something. Now when the voice in the burning bush said 'I am that I am,' it did not say 'I am love, I am truth.' Although later, Christ perhaps said that — that is, to start developing the meaning. But the primary meaning is, 'That I am, is what I am.' Or it could be translated as 'I am whatever I am,' but that's probably not quite so good as to say 'What I am is *that* I am' — in other words 'That I am is all.'

Q: This is really why I was looking at the meaning of the words and saying, well OK, if I accepted that as the meaning — 'I am that I am' — I may be confused with my depressive states or fearful states but I believe reality is love, unification and all that sort of stuff.

Q: You see, this emphasis on love still worries me because I would have thought that love came into planetary existence with man. If you go back two million years, or maybe two and a half million years, when you looked around the planet as we imagine it was — and it may be approximately something like the scientific picture — I think you would find it hard to say that this creation is the outcome of love and compassion. Blake said that love came with man. He said it poetically, but I would have thought that there was a case to suggest that love did, as it were, unfold with man and self-consciousness. Before then it's a creation all right; it's a fizzing, bubbling creation but I can't see much sign of love and compassion in it.

Q: Love has been defined as the urge in separated parts to re-unite. So the electron that has split off from the atom wants to come back. It's the same process at that level of coordination as would appear in the form of love between human beings at the human level of development. The inference is the flow of affinity which is the useful element at the heart of reality, which I think is very safe for development. It looks like love; that's one thing. But why are we speaking about God at all now, if we are speaking about reality? Because the God concept is only one way of pointing to reality. So why don't we talk about reality, and work our way from there?

Q: I second that motion.

Bohm: Yes. The reason we are talking about God is that people have talked about Him. It's part of the meaning of 'I am'. 'I am' has now been attached to the particular, and that is really the reality.

Q: We can discuss it as a program. We must become free from that.

Bohm: Yes. The difficulty is, does everybody regard it as a program?

Q: Probably most people would take the idea, the concept, as a program, as opposed to God as the complete self. That's not the right words, but what I'm trying to do is separate the concept of God from the idea of God itself. Does that make sense?

Q: Is the question you are asking: Does everyone see God as a program?

Bohm: Yes. You see, if we are going to communicate we have got to look at it together, right? You see, I'm not trying to propose answers here, but rather we have got to have communion.

Q: I feel that a predicated God, in some way a particularized form, must be a program. That's the way I perceive it. But I feel that the impulse behind it is something which isn't of the nature of a program. Whether one uses the phrase 'I am', or the word 'God', it seems to be the emergence of something into explicate form. It seems to be something of a much finer and subtler nature emerging, which we interfere with by making it into a program and defining it. It still symbolizes something else to me.

Bohm: It symbolizes a kind of creative explosion of energy, almost.

Q: Even the activation of energy in some way.

Q: I think the Jews had the right idea when the forbade the use of the name of God. *(laughter)*

Bohm: But they still did it indirectly.

Q: Could I come back just for a second to something we discussed yesterday, which was the idea that, at least from the point of view of quantum theory, there is no bottom line to reality. That would imply to me — I don't know whether this is a correct assumption or not — but that there is also no ceiling. Which means that the idea of a ground of all being as something that we can grasp or sink our teeth into, or use as a limiting concept, really is a mistaken notion.

Bohm: Yes. You see, this is what made me change this paper, because I discussed it originally in terms of mind and matter emerging from that kind of ground. Then I saw that yesterday we had discussed the soma-significance in which mind and matter just interpenetrate. So you can raise this question: Do we need this ground of all being? And that would be a very radical question, because it puts a very hard test on friendship, you see. *(laughter)* Which is more powerful, the friendship or the belief? We have to entertain this question.

Q: We have to discuss this because there is so much we don't know. And it's really nice to be able to trust this not knowing.

Bohm: Yes. I would say that we don't really know what's behind this, but what I want to call attention to is that these words with their meanings produce knowable effects. You see, we call attention to that. That is signa-somatic.

Q: We can have an insight into our intention in what we are doing, from what we are actually doing.

Bohm: And you can see that it shows the power of the ego being defined in that way. In a way, you could say that the ego tends to be identified with anything that is called or regarded as absolute. You see, the minute you think of the absolute, which has an unlimited signa-somatic power — the meaning of the absolute is unlimited, nothing stands before it, right? — the process of mind and body enacts this unlimited power within as the meaning. Acts it out. That power going on inside is then recognized by further thought as something extremely powerful but is not recognized as thought. You see there has been a failure of attention. Any power of that kind will interfere with attention. To release that much energy overloads the system; the blood pressure goes wrong, the chemicals go wrong, and so on. Therefore, in that inattention, the thought process attributes that enacted power of the unlimited to a being inside called 'me'. Because that would be the natural place where you would think that I am, right? Therefore you could say that the thought of the absolute will tend to create the thought of the ego if there is not absolute attention, whatever that would be. But whether such attention is even possible, is a question.

Q: I was thinking of an analogy. It's a little bit like living in a round room. Living in a round room is quite disturbing, because it doesn't have lines and corners that we usually have to limit our thoughts. And maybe the ego is something like that, in that we have to put down boundaries to identify ourself, and to stop ourselves from just spreading out into the universe, and letting our mind...

Bohm: Well, if the ego were acknowledged to be limited then it might serve a useful function. You see, we identify our self with the body and we say that our boundary is at the skin; but there's the well known example of a blind man tapping with a stick. If he holds the stick tightly, he feels that he ends at the end of the stick. If he holds it loosely, he ends at his fingers. You may regard the ego as in the head, or it may be that looking at your arm you don't see it as part of the ego, or you might identify the ego with a larger group and extend the boundaries. So the ego is a very ambiguous thing. It depends on what it means.

Q: Yes. Outside of that, if you cut off your leg, then

167

extend that further and further, and cut off your arm as well, and so on, would you still be here?

Bohm: Yes. Well, at some stage...*(laughter)* That emphasizes that the ego is defined by the meaning, which is ambiguous. Now it can be useful to identify the ego, up to a point, because every person has his own interests.

Q: What about James' definition, yesterday. He said it was really the spearhead. James, do you want to give your definition?

Q: I was giving the psychological description of the ego, which is the system which is built up by the mind to deal with the immediate environment. It's the sort of thrusting forward point that Jung pointed out. Ego is really one of the functions of mind, and the important thing is the self, which is the totality of the functions of mind — if we dare use such a word. But the ego is the instrument by which the unique individual, with his own bundle of eventualities, establishes him or herself within the environment.

Q: Can I enter in this? In any communication, a person, any person, could speak using the 'I' as coming from their archetypal self or their true essence. Or they could be coming from their limited self or ego. So every time any of us communicates to any other person using 'I', it leaves a sort of question mark about where they are coming from. In true communication it would always come from their true essence. But forgetfulness is continually occuring, and most of us, or some of us, speaking for myself, come from the ego, and that is missing the mark. So in a sense all egoic communication is less than real, less than true.

Q: I find there is some distinction between communion and communication, because in a deeper level there isn't the need for communication as much as there is the awareness of communion.

Bohm: Yes. That is a more fundamental...

Q: Are we not just as much defined by saying what we are not, as by saying what we are? One of the things we are programmed to say is we are not God, or we are not 'I am'. Therefore that in itself is a restriction. Therefore because we say we are not something, we have to say we are something else, and that causes...

Bohm: That complicates. But you see, also the very language says I am this. The meaning is there without our saying it, because it has been there so long in the human condition. Now you attempt to oppose that by saying 'I am not God.' But the notion that 'I am' is God is also implicit.

Q: But man as God on earth is particularly important in realizing what our lives on this planet are about. If we discard the ego, universalizing man to the extent that he is becoming a dimension of 'I am that I am', which is the force that we are representing, we deny the particular and then the very reason for our being here would cease to be.

Bohm: Yes. Well, we don't want to deny the particular, but we want to see that in some way there seems to be a danger in giving the value of the universal to the particular. In some way it may be correct. In some way the particular is an expression of the universal — an unfoldment — but the function of the program tends to make the particular the centre of everything, as if it were the universal 'I am'. You see, I think that's the danger. That is what I say is the source of egotism. Now it's a very delicate question because we don't want to wipe out the particular in trying to deal with this.

Q: The distinction between ego and egotism, I suppose, is quite clear to everyone. I think it is very important for this discussion. Ego is along the lines that James described it, which is the functional organization of the conscious part of one's world. Egotism is the identification of one's consciously accepted images and interests with the whole — that this is the paramount thing, and everything else should be subordinated to it. That distinction should be perfectly clear, so we can have this particular form of ego without giving in to the inflation and identification that comes through egotism.

Q: The ego says 'I am here.' Egotism says 'I am right.' *(laughter)*

Q: And incidentally, speaking of Moses, idolatry was the greatest sin. Idolatry, seen as a principle, seems crazy. Identification of the whole with a particular part — you take a bull or an object and you worship it as if it is God — but it isn't the whole. This is the greatest sin. This is fragmentation.

Q: Possibly the central problem of religion as a means of encompassing the whole, and the urge of the ego to be egotistical, comes from a pure perceptual fact. That is, I sit here in the centre of the universe of my perception. Everything surrounds me and therefore I come to think of everything as relating to me in the centre of my perceptual field. And religion somehow was contrived to counterbalance that. And so in fact, what's happened is that there is a tension that is never quite resolved between these two. But what strikes me is this word 'communion' which keeps coming up. Because once there is a communion, a real blending I suppose between two people, or a fusion of consciousness, the centre of that ego, the centre of that universe, moves from somewhere inside my skin to a new space between me and those who I'm in communion with. And the problem begins to soften, possibly.

Bohm: Yes. I think that what I was trying to say here is that we have got to see if we can dissolve this problem rather than resolve it.

Q: I'm just remembering that you spoke about everything

having meaning — the table, for instance — but that we've identified ourselves as the only objects which have this process of folding down from deeper levels of meaning, and folding back with soma-significant and signa-somatic, and in being conscious. It's almost as though the table has no problem with that — with having meaning. And in our consciousness, in my consciousness, I've become self-conscious; I mean self-conscious in the small sense of my meaning. And we've got this tool of thought — I don't know what we're meant to use it for — but this tool of thought which I apply in the wrong place. So I try with my thoughts to delve down into the depths of this well of meaning where it's perhaps inappropriate.

Q: Shouldn't one trust the meaning of the whole to move through a part and trust in being in the present moment? I don't know; I can't see it any other way.

Q: It's like I'm a predicating creature in some ways. It's a question of which way the attention is. If I'm going to predicate 'I am', then my attention is the wrong way. But the 'I am' will be predicated through me. It means I'm creating; there's nothing I can do to stop creating sub-wholes, and that is OK as long as my identification isn't in those sub-wholes. I have a responsibility for them, for the things I make manifest. But the 'I am' is not something I am trying to predicate. It will be predicated through me, and that then becomes my friend, my home, and so it's a question of where I'm looking. I accept that the way I think and act, in some way is meaning, and I will be either at a coarse or a subtle level depending on my relationship with the 'I am'. And as long as I accept that, I am going to create sub-wholes, but I don't want to treat that as me.

Q: In a way it's very simple. Our perspective should be wholeness.

Bohm: Yes.

171

Q: The intention is wholeness but we have to live as particulars.

Bohm: Yes. Well, we have to bring this all together. Now I think we could say that we are manifestations of the universal — each one of us — and we have to come in contact with this so that we can raise the question of whether we can combine the scientific and religious attitudes by beginning to inquire, not on the basis of knowledge, but by raising the question: Can wholeness be the subject of a free and unbiassed inquiry? Right? Which is both the interest of religion and science. We have to inquire. For example, we discussed that there may be a universal energy pervaded with intelligence and love which is the ground of everything — without belief and without disbelief. What I suggested was that saying the name of 'I am' signified this universal energy. Now is it possible for a human being, or a group of human beings, to actually come in contact with this universal energy, or be aware of that contact? Now this really is the question. If this is really possible then the egotism should go, right? I think this is really what religions are aiming for, and we have been discussing whether particular ways of putting religion would help or not. But I think that this is a very serious inquiry, and that we can't really settle the issues like that. But it's part of the evolution of human consciousness. That is, for example, the notion of God was originally rather limited, then came the notion of this universal God; but perhaps this notion has to develop or evolve further by inquiry.

Q: But the problem of inquiry into the whole is that you cannot conceptualize it.

Bohm: No. You can't.

Q: That would lead into paradox, and it does lead into paradox.

Bohm: Yes. The inquiry is whether it will lead to the dance of the soma-significance...

Q: That is critical, yes.

Bohm: ...which brings it to the whole.

Q: I'd like to add here that conceptualization might be only a means for us to get in touch with the whole, the universality. We can experience it with the whole of our being and not simply with our brain. And as we yield to our own nature more fully, we become aware of sensitivities which can make direct experience with the whole energy and process that is working through us. And through that we have more and more experience of the whole which encompasses. So conceptualization is a very poor, small part.

Bohm: Yes. I would only say that conceptualization has had a very big effect, which we have to give our attention to. Now you see, I think it's important that we should not in this inquiry define the situation so much as to disrupt the friendship. We have got to say that different people are approaching this in different ways, and each of us has got to respect the way that the other is approaching it, you see, because one may say that it's helpful to attribute qualities to God, which are limited, and that it helps as a metaphor, and others may say that it's not. Now I think that we begin by respecting each other's approach. In the spirit of friendship, we can then turn to the basic question — somehow we have got to come in contact with this beyond concept — of this universal energy, with our whole being.

You see, I have the feeling that when we started to define things too much, it was already beginning to disrupt, right? So in other words, that very process that I was talking about was beginning to happen. *(laughter)* Now that's not what we want, right? We don't really know the answer. There may be no answer. But it's really how this expresses itself in life. The question of the nature of 'I am' is really a crucial question, what it means in the whole of life. That's really the point to keep in mind.

Now we have this distinction of egotism and self, and the danger that this universal energy will be attached to the ego

173

and give it that meaning. I think that that really is the only
final point that I would want to make. People, in different
ways, are trying to approach this differently, and we have
got to respect each other. There are those who don't want
to give qualities to God, or those who don't even want to
say there is God, and there are those who do. Now this
would be an example of working together despite those
differences, because we realize that there is something
more important than that.

Q: Hurrah! *(laughter)*

Bohm: But we want to notice that this question has had a
powerful effect on human development, and how delicate
it is.

Q: I would like to really express my deep appreciation for
what you offered in these three days. There really has been a
deepening friendship, not only with you, but between all of
us. So I know that I'm beginning to appreciate even more
the unique qualities in each person, and how their processes
have been quite dominant in certain aspects, and how they
brought to point certain things that in me weren't as
dominant. I could say it complemented and enriched my
own experience because of that. And I think it's been a
beautiful unfolding of life. *(applause)*

Bohm: I wanted to say more or less the same thing.
(laughter) I've really learned a great deal here. As you see,
I had to give up my talk. *(laughter)* And I can see that
putting it in certain ways can have a disruptive effect, you
see, and I think we can learn from that, and that really
there is something much more real than all these questions,
and that is the actual way we are related, and what passes
between us. And we must not allow such questions to
break things up; but we've got to realize how they do
because that's part of the state of the world, right?

REMARKS ON THE PROCESS OF DIALOGUE

As mentioned in the introduction, the weekend began with the expectation that there would be a series of lectures and informative discussions with emphasis on content. It gradually emerged that something more important was actually involved — the awakening of the process of dialogue itself as a free flow of meaning among all the participants. In the beginning, people were expressing fixed positions, which they were tending to defend, but later it became clear that to maintain the feeling of friendship in the group was much more important than to hold any position. Such friendship has an impersonal quality in the sense that its establishment does not depend on a close personal relationship between participants. A new kind of mind thus begins to come into being which is based on the development of a common meaning that is constantly transforming in the process of the dialogue. People are no longer primarily in opposition, nor can they be said to be interacting, rather they are participating in this pool of common meaning which is capable of constant development and change. In this development the group has no pre-established purpose, though at each moment a purpose that is free to change may reveal itself. The group thus begins to engage in a new dynamic relationship in which no speaker is excluded, and in which no particular content is excluded. Thus far we have only begun to explore the possibilities of dialogue in the sense indicated here, but going further along these lines would open up the possibility of transforming not only the relationship between people, but even more, the very nature of consciousness in which these relationships arise.

DAVID BOHM.
London
February
1985

THE PARTICIPANTS

V.V. Alexander, London
Jean Bradbery, London
Volker Brendel, Rehovat, Israel
Monica Bryant, Brighton, Sussex
Guy Claxton, London
Leslie Cohen, Southampton, Hampshire
Gerda Cohen, Southampton, Hampshire
Joe Coulson, Southampton, Hampshire
Diana Durham, Chipping Campden, Gloucestershire
Karen Eyers, London
Morel Fourman, Ewelme, Oxfordshire
Adrianna Gheradini, Rome, Italy
Tom Guilbert, Lancaster, Lancashire
James Hemming, Teddington, Middlesex
Michael Hopwood, Guildford, Surrey
Alan Humphreys, Stockbridge, Hampshire
John Hunt, London
Bill Isaacs, Oxford, Oxfordshire
Chris Isbell, Eastleigh, Hampshire
Bernadette Kelly, Aberdeen, Scotland
Dick Kitto, Brightlingsea, Essex
Annette Leleur, Slagelse, Denmark
David Lesser, Mickleton, Gloucestershire

Joan Linley, Weybridge, Surrey
Tom Martinsen, Oslo, Norway
Alan Mayne, Milton Keynes, Buckinghamshire
Graham Phippen, Mickleton, Gloucestershire
Irene Prinsen, Amsterdam, Holland
Lida Radziwill, Rome, Italy
Marie-Louise Radziwill, Rome, Italy
Mike Robinson, London
Carol Rogers, California, U.S.A.
Tom Saunders, London
Michael Shaw, London
John Tomlinson, Chipping Campden, Gloucestershire
Suzette Van Hauen, Vedbaek, Denmark
Anne van de Zaande, Leiden, Holland
David Webb, Stockholm, Sweden
Joan Wells, Horsham, Sussex

COORDINATION

David Bohm, London
Sarah Bohm, London
Jenny Garrett, Mickleton, Gloucestershire
Peter Garrett, Mickleton, Gloucestershire
Anna Factor, Totnes, Devon
Donald Factor, Totnes, Devon

Printed in the United States
by Baker & Taylor Publisher Services

Printed in the United States
by Baker & Taylor Publisher Services